Luana Maria Alves Silva
Fernandes Almeida

Plant extracts in the management of Meloidogyne

Luana Maria Alves Silva
Fernandes Almeida

Plant extracts in the management of Meloidogyne

Peppers

ScienciaScripts

Imprint

Any brand names and product names mentioned in this book are subject to trademark, brand or patent protection and are trademarks or registered trademarks of their respective holders. The use of brand names, product names, common names, trade names, product descriptions etc. even without a particular marking in this work is in no way to be construed to mean that such names may be regarded as unrestricted in respect of trademark and brand protection legislation and could thus be used by anyone.

Cover image: www.ingimage.com

This book is a translation from the original published under ISBN 978-613-9-68187-7.

Publisher:
Sciencia Scripts
is a trademark of
Dodo Books Indian Ocean Ltd. and OmniScriptum S.R.L publishing group

120 High Road, East Finchley, London, N2 9ED, United Kingdom
Str. Armeneasca 28/1, office 1, Chisinau MD-2012, Republic of Moldova, Europe
Printed at: see last page
ISBN: 978-620-8-19472-7

SUMMARY

DEDICATE

To my parents Joaquim and Maria Araùjo, to my brothers Thiago Araùjo and Lili Alves, for encouraging me to achieve my goals, and who, even though they are far away, have always been there for me to make this dream come true!

I love you!

ACKNOWLEDGMENTS

To the Lord God who is with me every day.

To the Federal University of Piaui - UFPI-CPCE, especially to the Postgraduate Program in Agronomy - Plant Science, for the opportunity to study for a master's degree. To the PPGAF lecturers for their teachings over the years.

To the Coordination for the Improvement of Higher Education Personnel - CAPES, for the scholarship.

To my dear and wonderful parents, Joaquim Pereira da Silva and Maria Alves de Araùjo Silva, for their efforts, for their unconditional help, for raising and educating me so that I could achieve my goals, for their affection and encouragement, for always supporting me and supporting me even at a distance, with faith, courage and confidence in my potential. You are a fundamental part of my life, without you I would be nothing. I love you very much.

To my siblings, Thiago Araùjo and Lili Alves, for their affection and encouragement to always strive for the best.

To my advisor Prof. Dr. Fernandes Antonio de Almeida, I will carry with me your teachings, your friendship and my admiration, thank you very much for your respect, patience and guidance over the years.

To Prof. Dr. Jaime Maia, from the Phytosanitary Department of the State University of Sao Paulo - UNESP/Jaboticabal, for his support, patience, material support and lessons in nematology.

To my always dear Prof.ª Dr.ª Beatriz Meireles Barguil, for her countless teachings, contributions and friendship.

To my fellow undergraduates Thiago Pieta, Maisa Veras, Junior Prates, Paola Pieta, Fransley Félix, Ellen Cristina, Alexandre Martins, Rezanio Martins, Tabatha Reis, Betiara Dias, Keylane Mendes for helping me whenever I needed it while conducting my experiment.

To my great work partner Carmcm Abade, for her immeasurable friendship, companionship, availability, dedication, respect, professionalism and total support at every stage of this work.

To all my classmates in the Master's program in Agronomy-Phytotechny for their friendship and solidarity.

To the friends who, even at a distance, supported me, thank you for your companionship and affection.

Finally, to everyone who has always believed with me in my dreams, who has been there for me, who has been concerned, who has been supportive and who has cheered me on. Nothing in life is achieved alone, we always need other people to achieve our dreams and goals, so I am grateful to everyone who has contributed in one way or another to the completion of this work, my sincere thanks.

SUMMARY

The use of plants to extract plant-derived compounds used in phytonematoid management has gained emphasis as it represents another economically viable alternative with promising results. The aim of this study was to evaluate the potential of extracts from different plant species, the way they were prepared and the way they were applied to *Meloidogyne javanica*, applied via foliar spraying, soil addition and combined (soil + aerial part) to the pepper crop. In the first trial, the extracts were prepared in the Phytopathology laboratory using the cooking, aqueous and ethanolic methods. For the second trial, the plant extracts came from commercial formulations. The experiments were conducted in a greenhouse and in the Plant Pathology Laboratory. The evaluations consisted of plant height and stem diameter; fresh and dry phytomass of the aerial part of the plants; fresh and dry phytomass of the roots; root volume; number of galls and egg mass per 10g of roots and counting the number of juveniles in the soil of each treatment. In the first experiment, it was found that the extracts used showed satisfactory results in the management of phytonematoids compared to the control, being efficient in reducing the number of eggs, demonstrating potential in the management of nematodes, however the form of application of the extracts does not influence the reproductive characteristics of *Meloidogyne javanica* in the pepper crop. In the second experiment, the extracts with commercial formulations were effective in reducing the number of galls and juveniles in the soil, regardless of concentration. The extracts Pironat, Natunnem, Nematex at 1% and Rotenat and Natunnem at 10% showed the best results in reducing the number of galls. Rotenat at 10% applied to the soil and aerial part, Natunnem at 3% and 10% applied to the aerial part and combined and pyroligneous acid at 3% when applied via foliar spraying showed the best results in reducing egg masses. In view of the results obtained in the experiments, it can be concluded that the plant extracts made and the commercial formulations show potential in controlling phytonematodes, as well as being harmless to the applicator and the environment thanks to the natural origin of the substances involved.

1. GENERAL INTRODUCTION

The vast majority of economically important cultivated plant species are susceptible to attack by pests and diseases, which can cause productivity losses ranging from partial damage to total destruction (TIHOHOD, 2000).

Peppers (*Capsicum annuum* L.) are among the most widely grown vegetables in Brazil, serving different markets. The fruit is consumed *fresh,* as well as being used in the food industry to produce pigments (REIFSCHNEIDER & RIBEIRO, 2004). New production techniques, such as fertigation (MARCUSSI et al., 2004) and the use of hybrids have been applied, with the aim of improving quality and productivity, which was 290,000 tons in 2012 (CONAB, 2012).

With the inclusion of new technologies in the production system, various pests and diseases such as whitefly (*Bemisia tabaci* Genn.) (LIMA & CAMPOS, 2008), anthracnose (*Colletotrichum gloeosporioides* Penz.) (PARK, 2005), bacterial spot (*Xanthomonas vesicatoria* (Doidge) Dows) (LOPES, 2003) and galls (*Meloidogyne* spp.) (SOARES et al..) are extremely limiting and cause significant losses (*AZEVEDO,* 2006), 2006) are extremely limiting and cause significant losses to the crop (AZEVEDO, 2006).

Among the main biotic problems, phytonematodes have been found to directly interfere with the quality and productivity of the pepper crop (PINHEIRO et al., 2012). The most important nematodes are the so-called gall-formers which, when they penetrate the root, inject substances, giving rise to the so-called giant cells which provide them with food (FERRAZ, 2001).

In Brazil, of the *Meloidogyne* genus, the species *M. incognita* (Kofoid & White, 1919) Chitwood, 1949 and *M. javanica* (Treub, 1885) Chitwood, 1949, are considered the most aggressive, due to their high adaptability to the tropical region and their high pathogenicity for various economically important crops, such as vegetables (HUANG, 1992).

Most *C. annuum* cultivars are susceptible to gall nematodes, especially the Ikeda cultivar, which is widely cultivated in most Brazilian regions due to its ability to adapt to the climate and its high productivity rate (SANTOS, 2008).

The management of phytonematodes presents a series of inconveniences due to the high degree of polyphagism, parasitism and the diversity of host plants (SILVA et al., 2002). Among the options known and applied for control, nematicides or chemical products stand out, used on a large scale because they are highly efficient.

However, they cause a series of economic losses, as they considerably increase the cost of production (VIDA et al., 2004); interfere directly with the environmental microfauna (ANTES, 2008); and are highly toxic, exposing the applicator to the risk of poisoning (SILVA et al., 2002).

Currently, in Brazil, the main active ingredients registered with the Ministry of Agriculture, Livestock and Supply (MAPA) to control phytonematodes are restricted to a few crops, such as Aldicarb, which is indicated for use on cotton, potatoes, coffee, sugar cane, citrus fruits and beans, Carbufuran, indicated for cotton and tomatoes, and Abamectin, only for cotton (MAPA, 2012).

Due to the major problems already mentioned, the use of chemical products has been losing momentum for some years, especially among small and medium-sized producers in various countries (FERRAZ & VALLE, 2001).

In this way, various alternatives have been applied around the world for decades. Peres et al. (2003), in Spain, obtained satisfactory results when they evaluated the efficiency of vegetable oils in controlling gall nematodes, with a reduction of 83%. In Sudan, plant extracts from different plant species were used to control *M. incognita*, showing efficiency with more than 94% mortality of juveniles (ELBADRI et al., 2008). Bionematicides derived from plants are already used mainly in the United States, which demonstrates the expectations of their use in agriculture (FERRAZ et al., 2010).

Therefore, there are many lines of research employing pest and disease management, exploring the active principles of the most diverse forest species, applying the extract to the soil or incorporating the plant parts (CANNAYANE & RAJENDRAN, 2003;

LOPES, 2004) and foliar spraying (GARDIANO, 2006).

Plants have been used as another option in studies to control a wide variety of pathogens because they have metabolites with properties and substances present in the most diverse parts of the plant (CUNHA et al., 2003).

In view of the above, the aim of this study was to evaluate the efficiency of different plant extracts, preparation methods and application methods in controlling *M. javanica* in pepper crops.

1.1 BIBLIOGRAPHICAL REFERENCES

ANTES, V.P. **Parasitism** *of Meloidogyne* **spp. on native plants in western Paraná and genetic variability of populations of** *Meloidogyne incognita* **race 3**. 53f. 2008. Master's Degree. State University of Western Paraná.

AZEVEDO, C.P. **Epidemiology and control of anthracnose in** *Capsicum* **spp. and identification of** *Colletotrichum* **spp. associated with cultivated solanaceae**. 2006. 102 f. Thesis (Doctorate) - University of Brasilia, Brasilia, DF.

CANNAYANE, I.; RAJENDRAN, G. Penetration of *Meloidogyne incognita* (race 3) in tomato and brinjal roots treated with botanical extracts. **Indian Journal of Plant Protection**, v. 31, p. 84-86, 2003.

CONAB. Irrigation in pepper growing. 1st edition Brasilia, Ministério da Agricultura Pecuaria e Abastecimento, 2012. (**Circular tècnica 101**).

CUNHA, F.R.; OLIVEIRA, D.F.; CAMPOS, V.P. Plant extracts with nematicidal properties and purification of the active principle of *Leucaena deucocephala* extract. **Fitopatologia Brasileira**, Brasilia, n. 4, v.18, p.438441, 2003.

ELBADRI, G.A.; LEE, D.W.; PARK, J.C.; YU, H.B.; CHOO, H.Y. Evaluation of various plant extracts for their nematicidal efficacies against juveniles of *Meloidogyne incognita*. **Journal of Asia-Pacific Entomology**, Seoul, v.11, p.99-102, 2008.

FERRAZ, L.C.C.B. Soybean meloidoginosis: past, present and future. In. SILVA, J. F. V.; MAZAFFERA, P.; CARNEIRO, R.G.; ASMUS, G.L. FERRAZ, L.C.C.B.

Relações parasito-hospedeiro nas meloidoginoses da soja. Londrina, Embrapa Soja: Nematology Society, 2001. 127p.

FERRAZ, S.; FREITAS, L.G.; LOPES, E.A.; DIAS-ARIEIRA, C.R. **Manejo sustentável de Atonematoiodes**. Viçosa: UFV, 245 p. 2010.

FERRAZ, S.; VALLE, L. A.C. **Controle de fitonematóides por plantas antagonistas**, Viçosa: Editora UFV, 2001, 73p.

GARDIANO, C.G. **The nematicidal activity of aqueous extracts and plant tinctures on *Meloidogyne javanica* (Treub, 1885) Chitwood, 1949**, 2006. 92 f. Dissertation (Master's Degree) Universidade Federal de Viçosa, Viçosa, MG.

HUANG, S.P. **Nematodes that attack olericolas and their control**. Informe Agropecuârio, 1992, n. 172, v. 16, p. 31-36, Belo Horizonte.

LIMA, L.C.; CAMPOS, A.R. Factors affecting oviposition of Bemisia tabaci (Genn.) biotype B (Hemiptera: Aleyrodidae) on pepper. **Neotropica entomològica** [online], n.2, v.37, p.180-184, 2008. Available at   :

http://www.scielo.br/pdf/ne/v37n2/a12v37n2.pdf. Accessed on: Jan. 16, 2013.

LOPES, C.A.; AVILA, A.C. **Pepper diseases:** diagnosis and control. Embrapa Hortaliças. Brasilia, 2003, 96 p

LOPES, E.A. **Potential of aqueous extracts and soil incorporation of Mucuna Preta (*Mucuna pruriens* var. *utilis*) for the control of gall nematodes**. 2004 96 p. Dissertaçao (Mestrado) Universidade Federal de Viçosa, Viçosa, MG.

MAPA (Ministry of Agriculture, Livestock and Supply) / **Agrofit**.

Available   at: http://www.agricultura.gov.br/servicos-e-sistemas/sistemas/agrofit. Accessed on September 1, 2012.

MARCUSSI, F.F.N.; GODOY, L.J.G.; BÔAS, R.L.V. Nitrogen and potassium fertilization in pepper cultivation based on N and K accumulation by the plant. **Irrigaçâo**, Botucatu, n. 1, v. 9, p. 41-51, 2004.

PARK, S. K. **Differential interaction between pepper genotypes and**

Colletotrichum isolates causing anthracnose. 2005, 56 f. Dissertation (MSc) - Seoul National University, Seoul, Korea.

PÉREZ, M.P.; NAVAS-CORTÉS, J.A.; PASCUAL-VILLALOBOS, M.J.; CASTILLO, P. Nematicidal activity of essential oils and organic amendments from Asteraceae against root-knot nematodes. **Plant Pathology**. n. 3, v. 52, p. 395-401, 2003.

PINHEIRO, J.B.; AMARO, G.B.; PEREIRA, R.B. **Nematodes in chili peppers of the genus *Capsicum***. Ministry of Agriculture, Livestock and Food Supply (Circular Tecnica), 2012.

REIFSCHNEIDER, F.J.B.; RIBEIRO C.S.C. **Introduction and economic importance**. Brasilia: Embrapa Hortaliças, 2004. Available at: http://www.cnph.embrapa.br/sistprod/pimenta/index.htm. Accessed on: Jan. 15, 2013

SANTOS, P.V. **Reaçâo de acessos de pimenteiras (*Capsicum* spp.) a *Meloidogyne incognita* RAÇA 3**. 2008, 96f. Dissertation (Master's Degree) State University of Santa Cruz, BA.

SILVA, G.H.; OLIVEIRA, D.F.; CAMPOS, V.P. Purification of fungal metabolites with toxic effects on *Meloidogyne incognita*. **Fitopatologia Brasileira**, v.27, p.594-598, 2002.

SOARES, P.L.M.; BARBOSA, B.F.F.; BECARO, C.K.; GIMENES, R.; FERRAZ, M. P.S.; SANTOS, J.M.; BARBOSA, J.C.; MÙSCARI, A.M. Study of biological control of *Meloidogyne incognita* in the commercial production of 'Margarita' peppers in a protected environment. In. Brazilian Congress of Nematology, 2006, Rio de Janeiro. **Proceedings.** Brasilia: Brazilian Society of Nematology, 2006. p. 65. Abstract 18.

TIHOHOD, D. **Nematologia agricola aplicada**. 2 ed. Jaboticabal: FUNEP, 2000. 473p.

VIDA, J.B.; ZAMBOLIM, L.; TESSMAN, D.J.; BRANDÂO FILHO, T.; VERZIGNASSI, J.R.; CAIXETA, M.P. Management of Plant Diseases in Protected Cultivation. **Fitopatologia Brasileira**, n. 4, v. 29, p. 355-372, 2004.

CHAPTER I

PLANT EXTRACTS AND METHODS OF PREPARATION USED TO CONTROL MELOIDOGYNE JAVANICA IN CHILI PEPPERS

SUMMARY

SILVA, Luana Maria Alves, Federal University of Piaui, February 2013. **Plant extracts and methods of preparation used to control *Meloidogyne javanica* in peppers***. Supervisor: Prof. Dr. Fernandes Antonio de Almeida.

The search for alternative control methods aimed at preserving the environment has prompted research into plant extracts for phytonematoid control. The aim of this study was to evaluate the efficiency of different plant extracts, preparation methods and application methods in controlling *Meloidogyne javanica* in pepper crops. The experiment was conducted in a greenhouse and the Phytopathology Laboratory of the Federal University of Piaui, in Bom Jesus, PI. The experimental design was entirely randomized, in a factorial scheme (4 x 3 x 3 + 1), with the factors consisting of four plant extracts: Castor bean *(Ricinus Communis* L.), Caetanus melon *(Momordica charantia* L.), Cassava *(Manihotia charantia* L.).) cassava (*Manihot esculenta* Crantz) and neem (*Azadirachia indica* A. Juss), three methods of preparation (aqueous, cooking and ethanolic) and three forms of application (soil, aerial part and soil + aerial part), a control (water) with four replications. The pepper plants were inoculated with a suspension containing 5,000 eggs/juveniles of *M. javanica*, and after 72 hours 100 ml of the extracts were added to the soil, divided into two applications. Plant height, stem diameter, fresh and dry phytomass of the aerial part of the plants, fresh and dry phytomass of the root, root volume, number of galls, mass of eggs per 10g of roots and count of the number of juveniles in the soil of each treatment were evaluated 60 days after inoculation of the suspension. The extracts used showed satisfactory results in the management of phytonematoids in relation to the control, being efficient in reducing the number of eggs, demonstrating potential in the management of nematodes, however the form of application of the extracts does not influence the reproductive characteristics of *Meloidogyne javanica* in the pepper crop.

Keywords: Alternative control, gall nematode, *Capsicum annuum* L.

2.1 INTRODUCTION

Peppers (*Capsicum annuum* L.), which belong to the solanaceae family, are among the most widely cultivated vegetables in Brazil (FILGUEIRA, 2007). Annually, the area under cultivation is around 13,000 hectares, with production close to 290,000 tons of fruit. The largest producing states are Sao Paulo, Minas Gerais, Bahia and Rio de Janeiro (CONAB, 2012).

Most *C. annuum* cultivars are susceptible to gall-forming nematodes, such as the *M. javanica* species, showing a reproduction factor of 1.6, which makes it a good host (SANTOS, 2008). These results were also observed by Peixoto (1999), who when studying hybrids for their resistance to *Meloidogyne* spp., "Ikeda" stood out for its high susceptibility to nematodes of the same species mentioned above in the different races evaluated.

The management strategies used to control phytonematoids have not achieved satisfactory results in terms of prolonged control and lower risk of environmental impact. In particular, chemical products are among the most widespread control measures applied to nematodes, as they are highly efficient and produce quick results. However, they considerably increase the cost of production, as several applications are necessary, and even then they do not completely eliminate nematode populations (VIDA et al., 2004). Another negative aspect of the use of agrochemicals is the environmental damage, such as the presence of residues in the soil, water and air, in food which can compromise the quality of the water supply and the vegetables consumed, as well as the destruction of beneficial microorganisms in the soil (DORES & LAMONICA-FREIRE, 1999).

Due to the unfavorable effects of nematicides, alternative methods of control have been studied, such as the use of extracts from various species and parts of plants with nematicidal or nematostatic principles (FERRIS & ZHENG, 1999; NEVES et al.,

2005). When evaluating extracts from different plant species, Coimbra et al. (2006) observed the nematicidal effect of extracts from the leaves of manioc (*Manihot esculenta* Crantz), mint (*Mentha piperita* L.), garlic bulbs (*Allium sativum* L.), seeds and leaves of *papaya* (*Carica papaya* L.) and gliricidia bark (*Gliricidia sepium* (Jacq.) Steud.) on the nematode *S. bradys* (Stainer and LeHew). Neves et al. (2008) studied the nematicidal effect of papaya seed extract and found a 96.8% and 99.3% reduction in the hatching of *M. javanica* and *M. incognita* juveniles.

Extracts from plant species are promising for nematode management, especially for small planting areas. They have benefits when compared to chemical products, such as the ability to generate new compounds that pathogens are not yet able to inactivate, reduced toxicity, accelerated decomposition, a broad mode of action and being derived from reusable resources (COIMBRA et al., 2006).

Extracts can be prepared in different ways, and according to Martinez (2002), some factors such as the type of solvent and the extraction method used to acquire the extracts can influence the activity of the active ingredient. With regard to the forms of use, research has been carried out on extracts used in soil applications (CHITWOOD, 2002), which have shown good results, or via foliar application, which has not been studied much (GARDIANO et al., 2008a). Almeida et al. (2012) evaluated the efficiency of plant extracts and methods of preparation and found significant results in the management of gall nematodes in tomato crops in a greenhouse. The aqueous extract of neem leaves at doses of 1.5% and 3.0% used by Javed et al. (2008), when applied to the soil, led to reductions in the number of galls and eggs of *M. javanica*.

Substances such as azadirachtin, with proven efficiency, are found mainly in seeds and to a lesser extent in peels and leaves (MORDUE & NISBET, 2000). Coimbra et al. (2006) also proved the nematicidal effect of cassava leaves on the nematode *S. bradys,* causing 44.0% mortality of the population. This efficiency is due to the compounds found in its leaves, such as the cyanogenic glycoside, mainly Iinamarin, which when hydrolyzed releases cyanide gas, toxic to various microorganisms (NASU, 2008).

In this context, the aim of this study was to evaluate the effect of different methods of preparation and application of plant extracts for the management of *M. javanica* in the pepper crop in a greenhouse.

2.2 MATERIAL AND METHODS

The experiment was carried out in a greenhouse and the Phytopathology Laboratory at the Profª Cinobelina Elvas Campus of the Federal University of Piaui (UFPI) in Bom Jesus-PI.

The chemical analysis of the soil collected at a depth of 0-20 cm, before the experiment was set up, showed: pH (H2O): 6.9; calcium (Ca^2 +): 5 cmolc dm-3; aluminium (Al^3 +): 0 cmolc dm-^{3} ; magnesium (Mg^2 +): 2.5 cmolc dm^{-3} ; phosphorus (P, Mehlich method): 56.2 mg dm^{-3} and potassium (K^+): 187.3 mg dm .$^{-3}$

Soil treatment and seedling preparation

The soil used in the test was a mixture of soil-sand-manure, in the ratio 3:2:1 (v:v), respectively, previously treated with Dazomet (Basamid$^®$) at a dose of 50 g/m^2 , left to rest covered with a plastic tarpaulin for seven days to eliminate all possible microorganisms. After sterilization, the substrate remained exposed to the open air, being turned over and wetted, repeating this operation every 12 hours for three days. At the end, lettuce seeds (*Lactuca sativa* L.) were sown to assess whether the chemical product used in the sterilization still persisted. The use of this species is a consequence of its high sensitivity when exposed to the fumigant Dazomet. Finally, more than 98% germination was observed, demonstrating that the residual effect of the product had been satisfactorily degraded. This substrate was then distributed in polyethylene pots with a capacity of 8.0 liters, where 5 seeds of "Ikeda" pepper were sown. Thinning was carried out fifteen days after emergence, taking into account the vigor and height of the seedlings and leaving one plant per pot.

Obtaining and inoculating the pathogen

The inoculum was obtained from populations of *M. javanica* from pepper plantations in the municipality of Bom Jesus-PI. In order to neutralize any contamination in the test with other nematode species, the Coolen and D'Herde (1972) method was used to extract the nematodes from the roots, which consisted of grinding them with the aid of a blender for 15 seconds, at low speed, in a 0.5% sodium hypochlorite solution. The liquid was then poured through a 200 mesh sieve and collected in a plastic bucket; the suspension was then passed through a second 500 mesh sieve, discarding all the liquid. Using a pissette and strong jets of water, the nematodes retained on the 500 mesh sieve were collected in a 100 mL beaker, obtaining a final suspension of 40 mL, which was distributed into correctly balanced centrifuge tubes and centrifuged for 4 minutes at a speed of 2000 rpm. After the centrifugation, the supernatant liquid was removed by adding a sucrose solution (400g of sugar in 750mL of water) and another centrifugation was carried out for one minute at a similar speed to the first. At the end, the supernatant liquid was poured back into a 500-mesh sieve and the nematodes were collected in a 100mL beaker using water jets emitted by a pissette.

After extraction and identification based on the morphological characteristics of the species, inoculation was carried out on pepper plants grown in a greenhouse in treated soil, in order to multiply and obtain a pure suspension for use in the test. At the end of the multiplication period, the endoparasite specimens were extracted using the technique proposed by Hussey & Barker (1973), modified by Bonetti & Ferraz (1981), where a suspension (5,000 eggs/juvenile) of *M. javanica* contained in 10 mL of the suspension was obtained and calibrated using a Peters chamber under a stereoscopic microscope. The suspension was then taken to the greenhouse and inoculated into a single plant for each treatment, distributed in three holes 3.0 cm deep, 2.0 cm apart from each other and from the hypocotyl of the seedlings at 20 DAE (Days after emergence) in the soil contained in the pots. At the end of the inoculation, the plants were kept watered only twice a day for 72 hours to allow for parasitism.

Preparation of vegetable extracts

They were obtained from botanical material (leaves) of the plant species Mamona (*Ricinus communis* L.), Melâo-de-sâo-caetano (*Momordica charantia* L.) Cassava (*Manihot esculenta* Crantz) and Nim (*Azadirachia indica* A. Juss), all collected in the municipality of Bom Jesus-PI, in March 2012.

After collection, the leaves were dried at room temperature and weighed according to the method used to obtain the extracts, as follows:

Aqueous extracts

To prepare the extracts, 10 grams of dried leaves of the plant species were mixed with 100 mL of distilled water, left to stand for 24 hours and then filtered through Wathman paper n° 1. The extracts were placed in aluminum foil-covered erlenmeyer flasks, identified and used immediately afterwards, according to Ferris and Zheng (1999).

Cooking extracts

The same amount of plant material was used for the same volume of water, which was then heated to boiling point (100° C) for three minutes. The extracts were left to stand for 72 hours, then filtered through Wathman paper n° 1. They were stored in glassware (erlenmeyers), identified and covered with aluminium foil, and used immediately afterwards, according to the methodology described by Della Modesta et al. (1997).

Ethanolic extracts

We used 427 g of leaves from each species, which were dried at room temperature and then ground into 255 g of powder using a knife mill. The powdered material was macerated using 1000 mL of ethanolic solution (80%), which was filtered through Wathman filter paper n° 1. The solutions were distilled under pressure at an average temperature of 60 °C in a rotary evaporator to obtain 5 g of each crude ethanolic extract, according to the methodology described by Matos (1997). After obtaining the extracts in this form, a single dilution (13%) was used for all the plant extracts in the study.

Applications of plant extracts and evaluations

Alternating applications were made directly to the soil in the pots, the aerial part and together (soil and aerial part) using a hand sprayer. For this purpose, 100 mL of this solution was used for each treatment, divided into two applications at an interval of 15 days.

Sixty days after inoculation with *M. javanica* were evaluated: plant height and stem diameter using a ruler and caliper, fresh and dry phytomass of the aerial part of the plants cut at the base of the stalk and weighed using an analytical balance, dried in a forced ventilation oven at 60° C; fresh and dry root phytomass (after washing in running water to remove aggregates), root volume (calculated by displacing the volume of water in a beaker), number of galls and egg mass per 10g of roots and counting the number of juveniles in the soil of each treatment (JENKINS, 1964). Egg masses were counted using the 0.0015% Floxin-B staining method for 20 minutes (TAYLOR & SASSER, 1978).

The experimental design was entirely randomized, in a factorial scheme (4 x 3 x 3 + 1) and the factors consisted of four plant extracts: Castor bean, Melao-de-sao-caetano, Cassava and Neem, three modes of preparation (aqueous, cooking and ethanolic), three forms of application (soil, aerial part and soil + aerial part) and a control with water and four replications. The data was submitted to statistical analysis using the SISVAR statistical program. The results for number of galls, mass of eggs and juveniles in the soil were transformed to $\sqrt{x}$ +0.5. In the end, the means were compared using the Scott Knott test at a 5% probability level.

2.3 RESULTS AND DISCUSSION

The analysis of variance (Table 1) for the agronomic characteristics shows that there was no significant effect for the triple interaction extract x method of preparation x method of application. However, the extract x method of preparation interaction had a significant effect on all the variables studied, with the exception of stem diameter.

Table 1 - Summary of the analysis of variance for the sources of variation plant extracts (EXT), mode of preparation (MP) and forms of application (FA) for agronomic characteristics plant height (ALT), stem diameter (DIAM), aerial part dry phytomass (FSPA), root dry phytomass (FSR) and root volume (VR) in the pepper crop, UFPI-CPCE- 2012.

Sources of variation	GL	ALT	DIAM	FSPA	FSR	VR
EXT	4	20,07**	0,52 ns	7,40**	24,45**	23,42 **
MP	2	7,15**	0,20 ns	7,42**	19,55**	26,09**

FA	2	$1{,}76^{ns}$	$0{,}28^{ns}$	$0{,}50^{ns}$	$1{,}94^{ns}$	$3{,}59*$
Ext X MP	8	$2{,}70**$	$1{,}01^{ns}$	$2{,}27*$	$2{,}29*$	$3{,}03**$
Ext X FA	8	$0{,}72^{ns}$	$0{,}82^{ns}$	$1{,}12^{ns}$	$1{,}24^{ns}$	$1{,}34^{ns}$
FA X MP	4	$0{,}87^{ns}$	$1{,}79^{ns}$	$0{,}24^{ns}$	$0{,}49^{ns}$	$0{,}98^{ns}$
Interaction (Ext x MP x FA)	16	$1{,}21^{ns}$	$0{,}64^{ns}$	$1{,}11^{ns}$	$0{,}85^{ns}$	$0{,}83^{ns}$
Waste	132					
C.V (%)	-	10,43	17,78	23,99	30,86	26,51

* Significant at the 5% probability level, ** Significant at the 1% probability level and ns Not significant by the F test.

The method of preparation is considered to be one of the most important stages in the preparation of extracts, as it can directly interfere with their effect on plant pathogens. In particular, the time x temperature binomial used to extract the compounds present in the plant material. Some substances can volatilize or degrade when exposed to high temperatures, for example by cooking (NGUDI et al., 2003). According to Gruenwald et al. (2010), the active ingredient of a compound can behave differently depending on the part of the plant used. These statements corroborate Mazaro et al. (2008), who mention the presence of some substances in higher or lower concentrations of active principles in different ways of preparation.

All the extracts proved to be efficient for the different preparation methods, when compared to the control, promoting a significant increase in plant height (Figure 1). The method of preparation did not alter the efficiency of the cassava extract in promoting greater plant height. However, the aqueous neem extract prepared by cooking was less efficient in promoting the growth of pepper plants. The same behavior was observed for the cassava extract prepared by cooking. According to Martines (2002), the active principle of extracts can be influenced by factors such as the type of solvent and the way in which the extracts are extracted, and consequently their efficiency can be reduced.

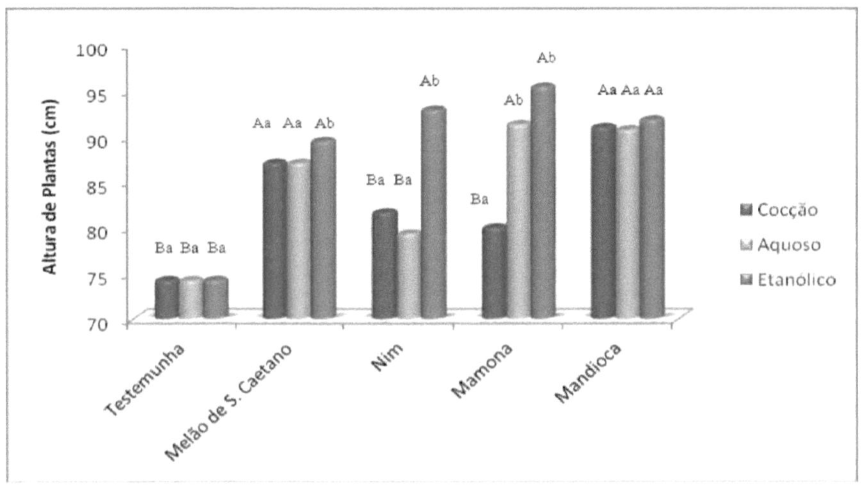

Means followed by the same letter in the columns and lower case in the row do not differ by the Scott Knott test at 1% probability.

Figura 1. Height of pepper plants inoculated with *Meloidogyne javanica*, treated with plant extracts in different forms of preparation in greenhouse conditions, Bom Jesus-PI, 2012.

For the dry phytomass of the aerial part (Figure 2), the best results were obtained by the extracts of São Caetano melon and cassava, when prepared by cooking. These results differ from those found by Gardiano et al. (2009), when they used the same extract in aqueous form to control *M. javanica* on tomato plants. Divergent results were also observed for cassava extract by Almeida et al. (2012), when they used the extract prepared by cooking. According to these authors, the method of preparation used may have contributed to the release of substances with allelopathic activity, making it impossible for tomato plants to grow. Plants such as cassava are rich in cyanogenic glucosides which, when degraded, can form aromatic aldehydes with allelopathic effects on various plant species (ALMEIDA, 1991). The good results obtained in this test may have been favored by several factors that can influence the active principle of the extracts, such as the age of the plant (POUTARAUD & GIRARDIN, 2005) the time of year when the plant material is collected for the preparation of extracts (PAPPIS et al., 2012); the part of the plant used, such as the leaf, root, seed or bark, which can have metabolite constituents that may or may not interfere as resistance inducers in plants when applied in extract form (GARDIANO et al., 2008a).

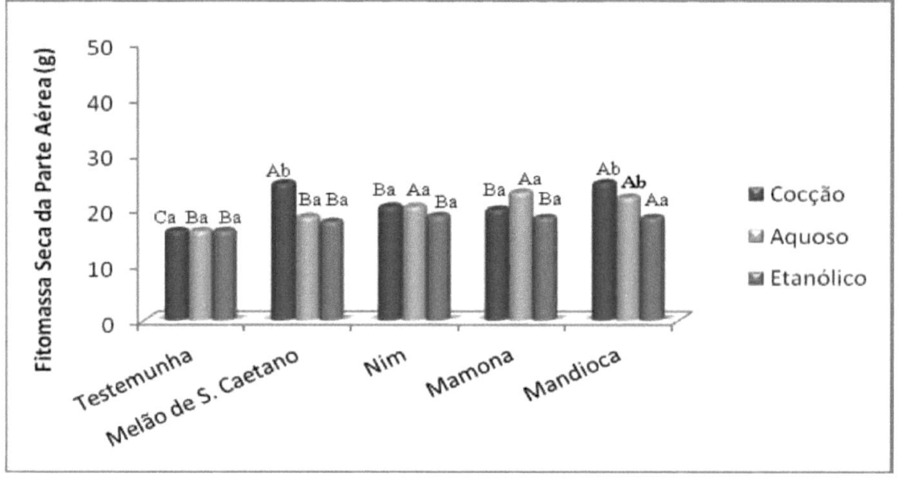

Averages followed by the same letter in the columns and lowercase in the row do not differ by the Scott Knott test at 5% probability.

Figura 2. Aerial part dry phytomass in pepper plants inoculated with *Meloidogyne javanica*, treated with plant extracts in different forms of preparation under greenhouse conditions. Bom Jesus-PI, 2012.

For root dry phytomass (Figure 3), all the extracts differed significantly when compared to the control. These results differ from those found by Gardiano et al. (2008a), who, when studying the effect of plant tinctures on *M. javanica* using neem extract, observed that these treatments did not significantly influence the weight of the root system. Mateus (2012) also observed no reduction in the root system with other extracts prepared in aqueous form.

The root system of plants is responsible for acquiring nutrients and H2O from the soil, and is of great importance for plant development. Therefore, a reduction in the development of the root system harms the plant's biological functions, making them more susceptible to water and nutritional deficiencies (GREGORY, 1994) and, consequently, favors the appearance of diseases (RODRIGUES et al., 2001).

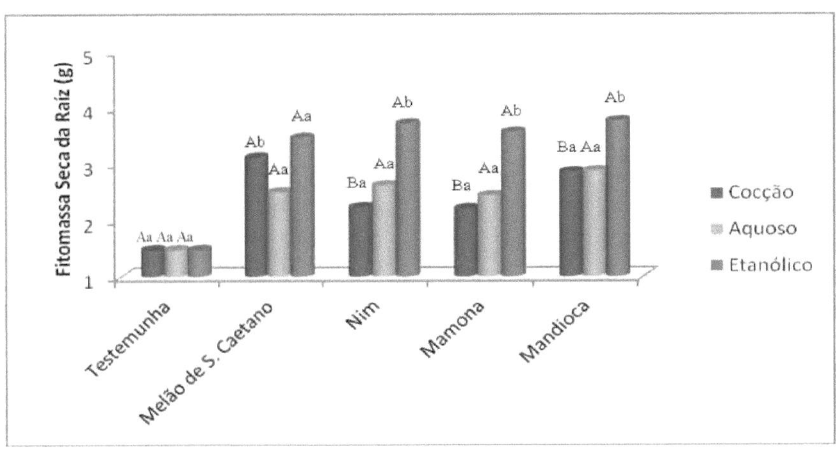

Averages followed by the same letter in the columns and lower case in the row do not differ by the Scott Knott test at 5% probability.

Figura 3. Root dry phytomass in pepper plants inoculated with *Meloidogyne javanica*, treated with plant extracts in different forms of preparation under greenhouse conditions. Bom Jesus-PI, 2012.

In Figure 4, all the extracts differed from the control in terms of root volume, which shows an increase in root volume in the plants treated with the extracts. This performance of the roots favors better nutrition for the plants, improving their health and helping them to resist various pests. The increase in the root system demonstrates the efficiency of the extracts in reducing nematode populations, favoring better root development (ZUCCHI, 2010).

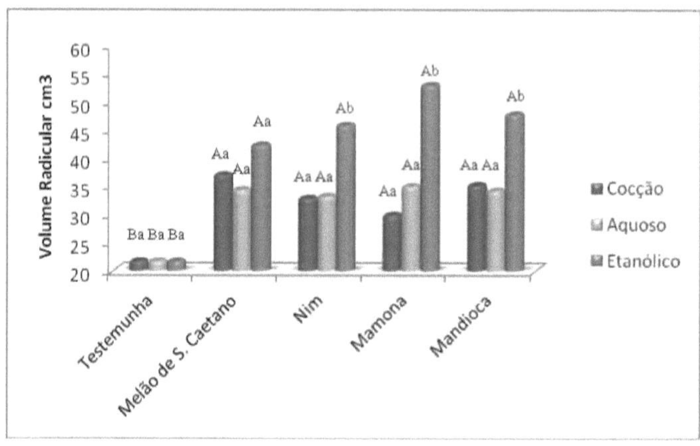

Means followed by the same letter in the columns and lower case in the row do not differ by the Scott Knott test at 1% probability.

Figura 4. Root volume in pepper plants inoculated with *Meloidogyne javanica*, treated with plant extracts in different forms of preparation under greenhouse conditions. Bom Jesus-PI, 2012.

The F values of the nematological characteristics for the different extracts, method of preparation and application are shown in Table 2. The plant extracts and the method of preparation had a significant influence on all the variables evaluated. At the same time, the results were different for the method of application, which did not affect any of the reproductive characteristics, thus not conferring a nematicidal or nematostatic effect. These results corroborate Gardiano (2006), who did not obtain significant results with foliar spraying of castor bean leaf extract, suggesting that this form of application is not suitable. Martines (2002) also found no positive results from foliar application of neem extract, one of the reasons being that the compound azadirachtin, present in the extract, is sensitive to sunlight and high temperatures, causing it to degrade and consequently reduce its activity.

There was a significant interaction between the factors for the number of galls, egg mass and number of juveniles, showing a broad influence of the extracts in reducing the population, and consequently the number of larvae ("F" values) (Table 2).

Table 2 - Summary of the analysis of variance (F values) of the factorial model for the sources of variation plant extracts (EXT), mode of preparation (MP) and forms of application (FA) for nematological evaluations number of galls (NG), number of egg masses (NMO), *Meloidogyne javanica* (MJ) in the pepper crop, UFPI- CPCE- 2012.

Sources of variation	GL	NG[1]	NMO[1]	MJ[1]
Extracts	4	142.72**	39,20**	135, 72**
Prep mode	2	8.32**	3,24*	79,18**
Form of application	2	0,95 ns	1,19 ns	1,95 ns
Ext X MP	8	2,36*	1,40 ns	5,17**
Ext X FA	8	0,54 ns	2,48**	3,19**
FA X MP	4	0,61 ns	0,95 ns	2,51*
Interaction (Ext x MP x FA)	16	1,22ns	0,84 ns	1,17 ns
Waste	132			
C.V (%)		23,78	18,14	17,06

Significant at 5% probability; ** Significant at 1% probability;ns Not significant. C.

V - coefficient of variation. [1] Original values transformed into square root of x + 0.5

For the number of galls variable (Figure 5), there was a reduction in all treatments compared to the control. The ethanolic extract of São Caetano melon obtained the best results, as did the cassava extract prepared by cooking. The reduction in the number of galls, in different quantities observed in the treatments, may be due to variations in the compounds and quantity of exudates released by the plant roots, which can undergo chemical reactions in the rhizosphere and degradation by microorganisms, as explained by Rocha & Campos (2004).

Martins et al. (2003), verified the efficiency of São Caetano melon extract in controlling *Rotylenchulus reniformis*. According to these authors, the presence of compounds in the leaves, such as alkaloids and momordic acid, prevented parasitism in the crop. Similar results with this plant species, used as an extract prepared by infusion, were observed by Dias et al. (2000) in the management of *M. incognita*, showing a percentage of dead juveniles of over 80%.

The results for the cassava leaf extracts corroborate Coimbra et al. (2006), who found the nematicidal effect of cassava leaf extracts and other vegetable extracts, causing mortality in the population of *Scutellonema bradys*, when using the cassava extract.

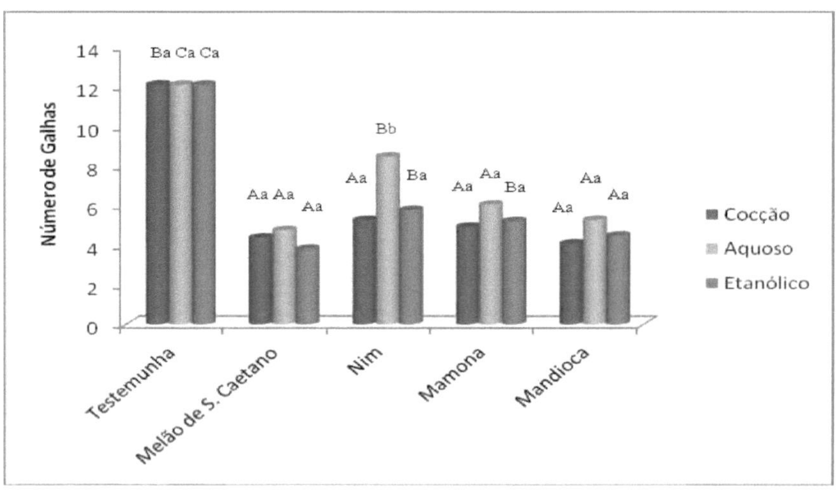

Averages followed by the same letter in the columns and lower case in the row do not differ by the test Scott knott at 5% probability

Figura 5. Number of *Meloidogyne javanica* galls on pepper plants treated with plant extracts in different forms of preparation under greenhouse conditions. Bom Jesus-PI, 2012.

For the number of egg masses (Figure 6), all the extracts differed from the control, confirming a nematicidal and nematostatic effect in all the forms applied. According to Vidhyaskaran (1992), the reduction in the number of egg masses may have occurred due to the activation of some plant defense mechanism, since the active ingredients present in extracts from plant species can act as elicitors and induce resistance in host plants, which will consequently reduce the development of the disease. Another explanation for the good results may be due to the exudates released by the roots of the plants treated with the extracts, or the presence of chemical components present in the extracts, absorbed by the plants and translocated to the roots, being released into the rhizosphere (ROVIRA, 1967).

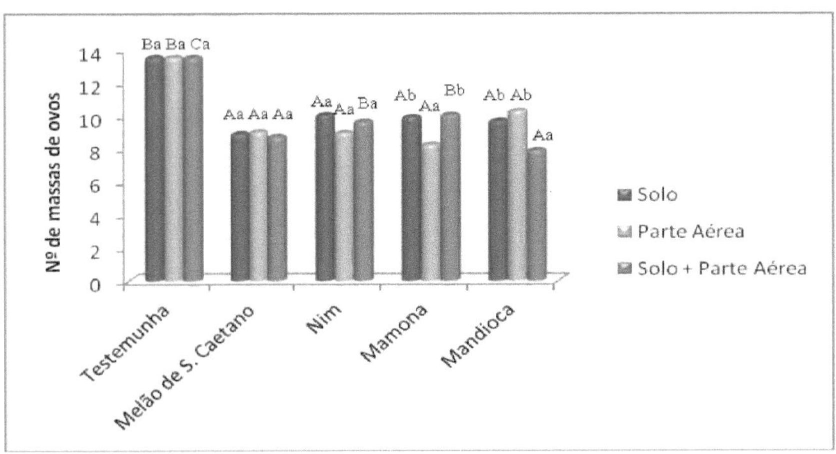

Averages followed by the same letter in the columns and lower case in the row do not differ by the Scott Knott test at 5% probability.

Figura 6. Number of egg masses of *Meloidogyne javanica*, in pepper plants treated with plant extracts, applied via soil, foliar spraying and combined in greenhouse conditions, Bom Jesus-PI, 2012.

For the number of *M. javanica* in the soil (Figure 7), the extracts prepared in different ways influenced parasitism, inhibiting its development and consequently its action on the roots. The extracts prepared by cooking and aqueous extracts stand out, even though they didn't differ from the ethanolic extract, they showed the best results. However, the lower results found in the plants treated with the ethanolic extract of São Caetanos melon when compared to the other forms of preparation may have been due to the non-release of the active ingredient by this form of preparation.

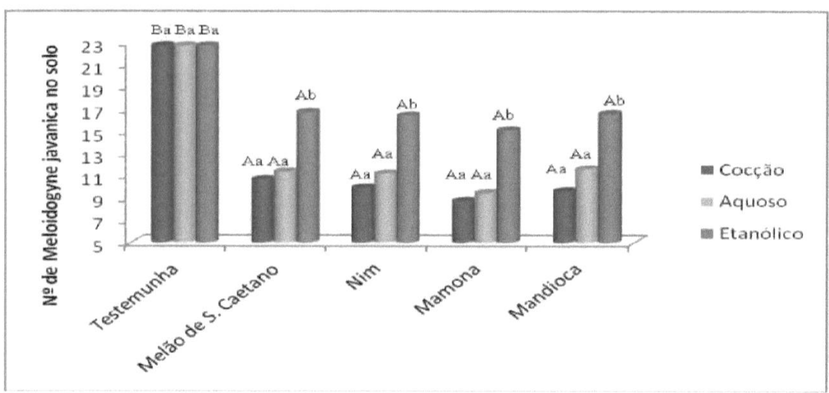

Averages followed by the same letter in the columns and lower case in the row do not differ by the Scott Knott test at 1% probability.

Figura 7. Number of *Meloidogyne javanica* in the soil, in pepper plants treated with plant extracts in different forms of preparation in greenhouse conditions, Bom Jesus-PI, 2012.

For the variable number of *M. javanica* in the soil, under different forms of application (Figure 8), all the extracts differed statistically when compared to the control. For the castor bean and cassava extracts, the results differed from the other extracts when applied to the soil and the aerial part. These results show the existence of organic compounds present in plants, called secondary metabolites, which, when synthesized by plants, exert pollinator attraction, environmental adaptation and phytoprotection activities (TAIZ & ZEIGHER, 2004), including against nematodes (FERRAZ et al., 2010).

26

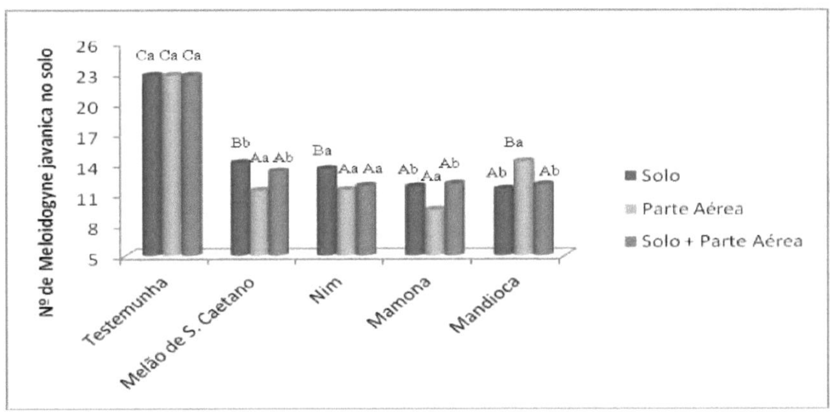

Averages followed by the same letter in the columns and lower case in the row do not differ by the Scott Knott test at 1% probability.

Figura 8. Number of *Meloidogyne javanica* in the soil, in pepper plants treated with plant extracts, applied via soil, foliar spraying and combined in greenhouse conditions. Bom Jesus-PI, 2012.

Also for the number of *M. javanica* in the soil, the interaction between application method and preparation method in Figure 9 showed that the best results were obtained when the extracts were applied to the soil in aqueous form and to the aerial part when the aqueous and ethanolic extracts were applied. Some studies have shown the efficiency of aqueous extracts in this form of application. Olabiyi (2008), when applying aqueous extracts obtained from the roots of the plants *Taguetes erecta* L., *Hypis suaveolens* (Poit) and *Ocimum gratissimum* L., observed a reduction in the reproduction rate and parasitism *of M. incognita*. Gardiano (2011), when evaluating the effect of aqueous extracts of plant species on the multiplication of *R. reniformis*, observed that the extract of Melâo de Sâo Caetano showed the greatest efficiency in controlling this nematode. According to Martins et al. (2003), the chemical constituents of Melâo de Sâo Caetano in its leaves are momordicine (alkaloid), momordicripine and momordic acid, and these compounds have nematicidal activity (DIAS et al., 2000). These results demonstrate the potential of the Melâo-Caetano extract as an alternative for controlling *M. javanica*.

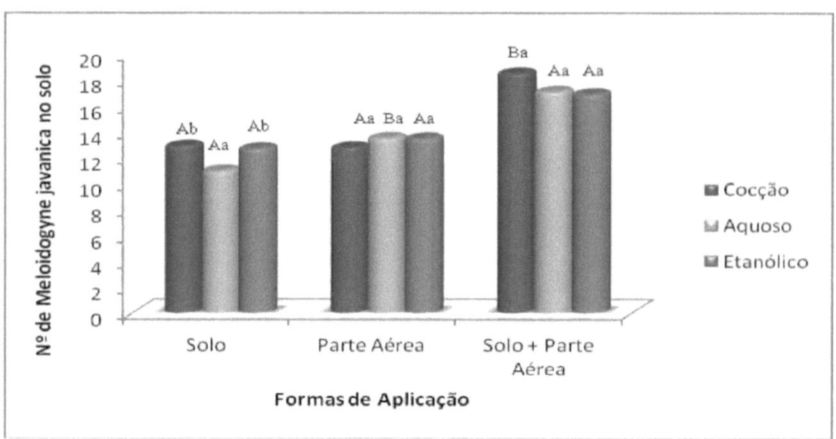

Averages followed by the same letter in the columns and lower case in the row do not differ by the Scott Knott test at 5% probability.

Figura 9. Number of *Meloidogyne javanica* in the soil, in pepper plants, in different forms of preparation, applied via soil, foliar spraying and combined in greenhouse conditions, Bom Jesus-PI, 2012.

2.4 CONCLUSIONS

All the extracts used showed satisfactory results in the management of phytonematodes compared to the control.

All the extracts in the different ways of preparation were efficient in reducing the number of eggs, showing potential in nematode management.

The way extracts are applied does not influence the reproductive characteristics of *Meloidogyne javanica* in pepper crops.

2.5 BIBLIOGRAPHICAL REFERENCES

ALMEIDA, F.S. Allelopathic effects of plant residues. **Pesquisa Agropecuâria Brasileira** v. 26, 221- 236, 1991.

ALMEIDA· F.A.; PETTER, F.A.; SIQUEIRA, V.C.; ALCÂNTARA NETO; F.; ALVES, A.U.; LEITE, M.L.T. Modes of preparation of plant extracts on *Meloidogyne javanica* in tomato. **Nematropica**, n. 1, v. 42, p. 9-15, 2012.

BONETI, J.I.S.; FERRAZ, S. Modification of the Hussey and Barker method for extracting *Meloidogyne exigua* eggs from coffee trees. **Fitopatologia Brasileira** v. 6, p. 553, 1981.

CHITWOOD, D.J. Phytochemical based strategies for nematode control. **Annual Review of Phytopathology,** v. 40, p. 221-249, 2002.

COIMBRA, J.L.; SOARES, A.C.F.; GARRIDO, M.S.; SOUSA, C.S.; RIBEIRO, F.L.B. Toxicity of plant extracts *to Scutellonema bradys*. **Pesquisa Agropecuâria Brasileira**, v. 41, p. 1209-1211, 2006.

CONAB. Irrigation in pepper growing. 1st edition Brasilia, **Ministério da Agricultura Pecuâria e Abastecimento**, 2012. (Circular tècnica 101).

COOLEN, W.C. D'HERDE, J. A method for the quantitative extraction of nematodes from plant tissue. Ghent, Belgian: **State of Nematology and Entomology Research Station**, 77 p. 1972.

DELLA MODESTA, R.C.; CARVALHO, J.L.V.; GONÇALVES, E.B.; VILLAMIL, C. I.V.; ALMEIDA, N.S.S. **Development of a sensory profile for Brazilian rice cultivars**. Rio de Janeiro: EMBRAPA-CTAA, 1997. 28 p. (EMBRAPA-CTAA. Boletim de Pesquisa, 21).

DIAS, C.R., A.V. SCHWAN, D.P. EZEQUIEL, M.C. SARMENTO, S.F. Effect of aqueous extracts of medicinal plants on the survival of juveniles *of Meloidogyne incognita*. **Nematologia Brasileira**, Brasilia, v. 24, p. 203-210, 2000.

DORES, E.F.; DE- LAMONICA-FREIRE, G.C. Ermelinda M. Contamination of the aquatic environment by pesticides: routes of contamination and dynamics of pesticides in the aquatic environment. **Journal of Ecotoxicology and the Environment (CEPPA).** Curitiba: Universidade Federal do Paranà, v.9, Jan/Dec, 1999.

FERRAZ, S.; FREITAS, L.G.; LOPES, E.A.; DIAS-ARIEIRA, C.R. **Manejo sustentàvel de fitonematoides**. Viçosa: UFV, 245 p. 2010.

FERRIZ, H.; ZHENG, L. Plant sources of chinese herbal remedies: effects on *Pra* tylenchus vulnus and *Meloidogyne javanica*. **Journal of nematology**, v. 31, n. 3, p.

241-263, 1999.

FILGUEIRA, F.A.R. **Novo manual de olericultura**: Agrotecnologia moderna na produção e comercializaçao de hortaliças. 3 ed. Viçosa: UFV, 2007. 402p.

GARDIANO, C.G. **The nematicidal activity of aqueous extracts and plant tinctures on *Meloidogyne javanica.*** 92p; Dissertation (Master's Degree). Federal University of Viçosa, 2006.

GARDIANO, C.G.; S. FERRAZ, E.A. LOPES, P.A. FERREIRA, S.C. Spraying vegetable dyes on tomato plants to control *Meloidogyne javanica*. **Revista Tròpica - Ciências Agràrias e Biológicas**, v. 2, p. 22-27, 2008a.

GARDIANO, C.G.S.; FERRAZ, E.A. LOPES, P.A. FERREIRA, D.X. AMORA, L.G.F. Evaluation of aqueous extracts of various plant species, applied to the soil, on *Meloidogyne javanica* (Treub, 1885) Chitwood, 1949. **Semina: Ciências Agràrias** v. 30, p. 551-556, 2009.

GARDIANO, C.G.; FERRAZ, S.; LOPES, E.A.; FERREIRA, P. A.; CARVALHO, S.L.; FREITAS, L.G. Evaluation of Aqueous Extracts of Plant Species, Applied via Foliar Spraying, on *Meloidogyne javanica*. ***Summa Phytopathologica***, Botucatu, n. 4, v. 34, p. 376-377, 2008b.

GARDIANO, C. G., MURAMOTO, S.P., KRZYZANOWISKI, A.A., ALMEIDA, W.P.; SAAB, O.J.G.A. Effect of aqueous extracts of plant species on the multiplication of *Rotylenchulus reniformis* Linford & Oliveira. **Instituto Biològico**, Sao Paulo, n. 4, v.78, p.553-556, 2011.

GOVINDACHARI, T.R.; SURESH, G.; GOPALAKRISHNAM, G.; BANUMATHY, B.; MASILAMANI, S. Identification of antifungal compounds from the seed oil of *Azadirachta indica*. **Phytoparasitica,** New York, n. 2, v. 26, p. 1-8, 1998.

GREGORY, P.J. Root growth and activity. In: **Phisiology and determination of crop yield**.chap. 4, p. 65-93, 1994.

GRUENWALD, J.; FREDER, J.; ARMBRUESTER, N. Cinnamon and Health. **Critical Reviews in Food Science and Nutrition**, v. 50, p. 822-834, 2010.

JAVED, N.; GOWEN, S.R.; INAM-UL-HAQ, M.; ABDULLAH, K.; SHAHINA, F. A Persistent effect of neem (Azadirachta indica) end formulations against root-knot nematodes, *Meloidogyne javanica* and their storage life. **Crop Protection**, v. 26, p. 911-916, 2008.

JENKINS, W.R.; TAYLOR, D.P. Chemical control In. Plant nematology, Ed. Peter **Gray Reinhold publishing corporation**, New York, 222-223 p, 1967.

MARTINEZ, M.M. Action of neem on nematodes. In: **Neem - *Azadirachta indica*, nature, multiple uses and production**. IAPAR Instituto Agronòmico do Paranâ, Londrina-PR. p. 65-68, 2002.

MARTINS, E.R.; CASTRO, D.M.; CASTELLANI, D.C.; DIAS, J.E. **Plantas medicinais**. Viçosa, MG: Editora UFV, 2003.

MATEUS, M.A.F. Aqueous extracts of medicinal plants in the control of gall nematodes. 2012. 59f. **Dissertation (Master's Degree)** Midwestern State University.

MATOS, F.J.A. **Introdução a phitoquimica experimental**. 2.ed. Fortaleza: UFC, 141 p. 1997.

MAZARO, S.M., CITADIN, I.; GOUVÊA, A.; LUCKMANN, D.; GUIMARÂES, S.S. Induction of phytoalexins in soybean cotyledons in response to derivatives of pitangueira leaves. **Ciência Rural** v. 38, p. 1824-1829, 2008.

MORDUE, A.J.; NISBET, A.J. Azadirachtin from the neem tree *Azadirachata indica*: its action against insects. **Anais da Sociedade Entomológica do Brasil**. v.29, p.615-632, 2000.

NASU, E.G.C. **Chemical composition of manipueira and its potential for controlling *Meloidogyne incognita* in tomato plants in western Paranà**. Dissertation (Master's Degree) - State University of Western Paranà, Marechal Cândido Rondon Campus. 56f, 2008.

NEVES, W.S.; FREITAS, L.G.; DALLEMOLE-GIARETTA, R.; FABRY, C.F.S.; COUTINHO, M.M.; DHINGRA, O.D.; FERRAZ, S.; DEMUNER, J.A. Activity of garlic (*Allium sativum*), mustard (*Brassica campestris*) and chili pepper (*Capsicum frutescens*) extracts on the hatching of *Meloidogyne javanica* juveniles. **Nematologia Brasilieira**, v. 29, p. 273-278, 2005.

NEVES, W.S.; FREITAS, L.G.; FABRY, C.F.S.; DALLEMOLE-GIARETTA, R.; FERREIRA, P.A.; FERRAZ, L. O.; DHINGRA, O.D.; FERRAZ, S. Nematicidal Action of Oil, Plant Extracts and Two Products Based on Capsaicin, Capsinoids and Allyl Isothiocyanate on Juveniles of *Meloidogyne javanica* (Treub) Chitwood. **Nematologia Brasileira**, n. 2, v. 32, 2008.

NGUDI, D.D., Y.H. KUO, F.L. Cassava cyanogens and free amino acids in raw and cooked leaves. **Food and Chemical Toxicology, v.** 41, p. 1193-1197, 2003.

OLABIYI, T.I. Pathogenicity study and nematotoxic properties of some plant extracts on the root-knot nematode pest of tomato, Lycopersicon esculentum (L.). **Plant Pathology Journal**, n. 1, v. 7, p. 45-49, 2008.

PAPPIS, L.; GINDRI, A.L.; PIANA, M.; ZADRA, M.; SOUSA, L.B.; ATHAYDE, M. L. MARIO, D.N. ALVES, S.H. Analysis of the antimicrobial activity of the crude extract and fractions of the leaves of *Urera bacifera* Gaudich (Urticaceae). **Revista de fitoterapia.** V. 12, p. 172, 2012.

PEIXOTO, J.R.; MALUF, W.R.; CAMPOS, V.P. Avaliação de Linhagens, Hibridos F1 e Cultivares de pimentâo quanto à resistência a *Meloidogyne* spp. **Pesquisa agropecuâria Brasileira**, Brasilia, n. 12, v.34, p.2259-2265, 1999.

POUTARAUD, A.; GIRARDIN, P. Improvement of medicinal plant quality: a Hypericum perforatum literature review as an example. **Plant Genetic Resources**, Cambridge, n. 2, v. 3, p. 178-189, 2005.

ROCHA, F.S.; CAMPOS, V.P. Effect of Plant Cell Culture Exudates on Second-Stage Juveniles of *Meloidogyne incognita*. **Fitopatologia brasileira**. n. 3, v. 29, May - Jun 2004.

RODRIGUES, F.; DATNOFF, L.E; KORNDORFER, G.H.; SEEBOLD, K.W.; RUSH, M.C. Effect of silicon and host resistance on sheath blight development in rice. **Plant Disease**, n. 8, v.85, p.827-832, 2001.

ROVIRA, A.D. Plant Root Exudates. **Botanical Review**, p. 35 - 53, 1967.

SANTOS, P.V. Reaçao de acessos de pimenteiras (*Capsicum* spp.) a Meloidogyne incognita RAÇA 3. 2008, 96f. Dissertaçao (Master's Degree) Universidade Estadual de Santa Cruz, BA.

TAIZ, I.; ZEIGER, E. Secondary metabolites and plant defense. In: TAIZ, I.; ZEIGER, (Eds.) **Plant physiology**. Porto Alegre, RS; Artimed, p. 309-332, 2004.

TAYLOR, A.L.; SASSER, J.N. Biology, identification and control of root-knot nematodes (*Meloidogyne* species). Raleigh: **International Meloidogyne Project**, NCSU & USAID Coop. 111p, 1978.

VIDA, J.B.; ZAMBOLIM, L.; TESSMAN, D.J.; BRANDÂO FILHO, T.; VERZIGNASSI, J.R.; CAIXETA, M.P. Management of Plant Diseases in Protected Cultivation. **Fitopatologia Brasileira**, n. 4, v. 29, p. 355-372, 2004.

VIDHYASKARAN, P. Principles of Plant Pathology. New Delhi: **CBS Printer and Publishers.** 1992.

ZUCCHI, F. *Trichoderma* in cultivated areas. **JV Biotecnologia**, 2010.

CHAPTER II

EVALUATION OF COMMERCIAL EXTRACTS IN THE MANAGEMENT *OF* MELOIDOGYNE JAVANICA *ON PEPPERS*

SUMMARY

SILVA, Luana Maria Alves, Federal University of Piaui, February 2013. **Evaluation of commercial extracts in the management of *Meloidogyne javanica* in peppers***. Supervisor: Prof. Dr. Fernandes Antonio de Almeida.

In the search for alternatives in the management of *Meloidogyne javanica* in the pepper crop, the aim of this study was to evaluate the effect of different concentrations and forms of application of commercial extracts in the management of gall nematodes in peppers. The experiment was conducted in the greenhouse and phytopathology laboratory of the Federal University of Piaui, in Bom Jesus. The experimental design was entirely randomized, in a factorial scheme (5 x 3 x 3 + 1), and the factors consisted of five commercial formulations based on plant species: Pironat® (pyroligneous acid), Rotenat® (rotenone), Natunnem®, Nematex® (azadirachtin) and crude pyroligneous extract, three concentrations (1%, 3% and 10%), and three forms of application (soil, aerial part and soil + aerial part) in four replications, and a control (water). The pepper plants were inoculated with a suspension containing 5,000 eggs/juveniles of *M. javanica* and left without any control measures for 72 hours. At the end of this period, 100 ml of the extracts were added to the soil, divided into two applications. At 60 days after inoculating the suspension with nematodes, plant height, stem diameter, fresh and dry phytomass of the aerial part of the plants, fresh and dry phytomass of the roots, root volume, number of galls, egg mass per 10g of roots and count of the number of juveniles in the soil of each treatment were evaluated. The extracts with commercial formulations were efficient in reducing the number of galls and juveniles in the soil, regardless of concentration. The extracts Pironat, Natunnem, Nematex at 1% and Rotenat and Natunnem at 10% showed the best results in reducing the number of galls. Rotenat at 10% applied to the soil and to the aerial part, Natunnem at 3% and 10% applied to the aerial part and combined and pyroligneous acid at 3% when applied via foliar spraying showed the best results in reducing egg masses.

Keywords: *Capsicum annuum* L., organic nematicides, gall nematode.

3.1 INTRODUCTION

The damage caused by phytonematodes, especially in economically important crops such as tomatoes (CAMPOS, 2000), lettuce (WILCKEN et al., 2005) and peppers (KUROZAWA et al., 2005), has caused significant production losses in modern agriculture (EMBRAPA, 2006).

In Brazil, the area under pepper cultivation is approximately 13,000 hectares, with Sao Paulo, Minas Gerais, Bahia and Rio de Janeiro as the main producing states (CONAB, 2012). However, in the presence of nematodes, crop production can be considerably reduced, making it a limiting factor for cultivation (KUROZAWA et al., 2005).

Control methods that reduce nematode population densities to levels that do not cause economic damage have been sought by many producers facing this problem. Alternative methods are currently being sought to replace or reduce the use of pesticides in agriculture (DOIHARA, 2005), because although nematicides are effective in controlling pathogenic nematodes, they are harmful to humans and the environment and, for the most part, are expensive, making them unviable for small farmers (MOHAN & SUBHASHINI, 2010).

In recent years, the use of plant parts in the management of plant-parasitic nematodes has been of great importance as a promising source of biopesticides, due to greater environmental awareness and knowledge of the risks to human health associated with chemical nematicides (SIVAKUMAR & GUNASEKARAN, 2011).

The use of natural products with commercial formulations has been widely used to control pests and diseases in agriculture (AZEVEDO et al., 2005). These products have advantages over chemical products, such as: favoring a suitable environment for the development of beneficial microorganisms, influencing disease prevention (NATURAL RURAL, 2010), they can be used directly by the farmer without risk to the applicator (GARDIANO, 2006) and they are also biodegradable (COIMBRA et al.,

2006).

The aim of this study was to evaluate the effect of different concentrations and ways of applying commercial extracts on the management of *Meloidogyne javanica* in peppers.

3.2 MATERIAL AND METHODS

The experiment was carried out at the Federal University of Piaui (UFPI) in Bom Jesus-PI, in the Phytopathology laboratory and in a greenhouse in the UFPI-CPCE experimental area.

The chemical analysis of the soil samples collected at a depth of 0-20 cm, before the experiment was set up, showed: pH (H2O): 6.9; calcium (Ca^{2+}): 5 cmolc dm^{-3} ; aluminium (Al^3 +): 0 cmolc dm^{-3} ; magnesium (Mg^{2+}): 2.5 cmolc dm^{-3} ; phosphorus (P, Mehlich method): 56.2 mg dm^{-3} and potassium (K+): 187.3 mg dm .$^{-3}$

The experimental design was entirely randomized, in a factorial scheme (5 x 3 x 3 + 1), and the factors consisted of five commercial formulations based on plant species: Pironat® (Pyroligneous Acid), Rotenat® (Rotenone), Natunnem® , Nematex® (azadirachtin) and crude pyroligneous extract, three concentrations (1%, 3% and 10%) and three forms of application (soil, aerial part and soil + aerial part) in four replications, and a control with water only. The data was submitted to analysis of variance and the means were compared using the Scott-Knott test at 5% probability.

The soil used in the test consisted of a mixture of soil-sand-dirt, in the ratio 3:2:1 (v:v), previously autoclaved at 120°C for one hour and then distributed in polyethylene pots with a capacity of 8.0 dm^{-3} , where 5 "Ikeda" pepper seeds were sown, and thinning was carried out 15 days after emergence, taking into account their vigor and height, leaving only one plant per pot.

The inoculum used in the study was obtained from populations of *M. javanica* from plantation areas in the municipality of Bom Jesus-PI. To this end, the Coolen & D'Herde (1972) method was used to extract juveniles which were inoculated onto pepper plants kept in a greenhouse for multiplication and isolation of the species to be used. After the species had multiplied, the juveniles and eggs were extracted according to the methodology described by Hussey & Barker (1973), modified by Bonetti & Ferraz (1981), in which the pepper roots were crushed in a 0.5% sodium hypochlorite solution for 15 seconds at low speed. The solution obtained was collected and transferred to a beaker, and calibrated to a concentration of approximately 5000 eggs/juvenile in 10 mL of the suspension using the counting chamber (Peters Chamber) through a stereoscopic microscope.

Forty days after sowing, the soil was infested with a suspension of eggs and juveniles, inoculated using a pipette and distributed in three holes about 3.0 cm deep in the soil, close to the root system of the plants. In the first 72 hours after inoculation, the plants were carefully watered to prevent loss of the inoculum through leaching and thus provide all the ideal environmental conditions for the phytonematodes, facilitating parasitism in the roots.

After the period of facilitating plant infection, two applications of the different extracts were made at 15-day intervals. The applications consisted of soil, aerial part and soil plus aerial part with a volume of 100 mL for each treatment. For the control, only water was used.

Sixty days after the inoculation of the suspension containing *M. javanica* eggs, the following variables were evaluated: plant height and diameter, fresh and dry phytomass of the aerial part; fresh and dry phytomass of the root, root volume (calculated by displacing the volume of water), number of galls and egg mass per 10g of roots, and counting the number of juveniles in the soil of each treatment (OOSTENBRINK, 1966). To count the egg masses, the method used was staining with 0.0015% Floxin-B for 20 minutes, according to the method described by Taylor & Sasser (1978).

3.3 RESULTS AND DISCUSSION

In the analysis of variance of the data for the different parameters evaluated (Table 1), with the exception of the form of application, there was a significant difference for all the agronomic characteristics. Thus, the direct influence of plant extracts on phytonematoid control was noticeable. There were double interactions for all variables and triple interactions for stem diameter and root dry phytomass. The results demonstrate the interdependence of the factors evaluated.

Table 1 - Summary of the analysis of variance of the factorial model for the sources of variation plant extracts (EXT), concentrations (C) and forms of application (FA) for agronomic characteristics plant height (ALT), stem diameter (DIAM), aerial part dry phytomass (FSPA), root dry phytomass (FSR) and root volume (VR) in the pepper crop, UFPI-CPCE- 2012.

Sources of variation	GL	ALT	DIAM	FSPA	FSR	VR
EXT	5	6,16**	7,29**	12,37**	4,19**	3,61**
C	2	5,05**	2,79 ns	9,70**	1,32 ns	0,81 ns
FA	2	3,95 ns	0,57 ns	1,16 ns	0,17 ns	0,45 ns
Ext X C	10	3,31**	6,04**	6,22**	4,03**	3,04**
Ext X FA	10	1,35 ns	5,20**	2,46**	1,45 ns	1,58 ns
FA X C	4	0,49 ns	1,46 ns	0,83 ns	1,23 ns	1,63 ns
Interaction (Ext x C x FA)	20	0,55 ns	2,98**	1,10 ns	1,90**	0,81 ns
Waste	159					
C.V (%)		11.32	8.36	19.50	30.52	33.17

Significant at 5% probability; ** Significant at 1% probability; ns Not significant. C. V - coefficient of variation

For the agronomic variable plant height, there was a significant difference for all the extracts applied at the 1% concentration and at all concentrations for the nematex and pyroligneous acid extracts. At 10%, the pyronat, nematex and pyroligneous acid extracts were superior, showing the best results for this parameter (Table 2). Akhtar & Mahmood (1996) demonstrated the efficiency of industrialized products based on plant extracts containing neem triterpenes (Nimin®) in controlling *M. incognita* in pepper plants, providing significant results in reducing the number of galls and consequently promoting plant growth, since the extracts can have a tonic effect on the vegetative development of the crop (FRANZENER, et. al., 2007).

Positive results with the use of commercial extracts have also been found in the control of agricultural pests. In the management of *Bemisia tabaci* biotype b (Hemiptera:

Aleyrodidae) on melon trees, the use of neem oil and pyroligneous extract caused a reduction of 66.49% and 67.35%, respectively (AZEVEDO et al., 2005).

The good results found for the control of pests and diseases in agriculture have been known for a long time, thanks to the composition of these products, such as monoterpenes, sesquiterpenes and flavonoids with pesticidal and efficient actions for the control of various phytopathogens (SCHMUTTERER, 1990).

Table 2. Plant height of pepper plants inoculated with *Meloidogyne javanica*, treated with plant extracts in different concentrations under greenhouse conditions, Bom Jesus-PI, 2012.

	Plant height		
	Concentrations		
Extracts	1%	3%	10%
Witness	87.00 Ba	87.00 Ba	87.00 Ba
Rotenat	96.91 Aa	90.91 Ba	89.16 Ba
Pironat	95.58 Aa	89.33 Ba	96.75 Aa
Natunnem	96.66 Ab	89.34 Bb	71.75 Ca
Nematex	93.50 Aa	94.99 Aa	96.50 Aa
Pyrolytic acid	98.00 Aa	95.83 Aa	93.41 Aa
C. V. = 11.32			

Averages followed by the same capital letter in the column and small letter in the row do not differ by the Scott-Knott test at 1% probability.

For the triple interaction of the stem diameter parameter (Table 3), there was no positive difference between the extracts and the control, except for 10% Nematex, when applied to the aerial part, which induced an increase of 25% and 65% in stem diameter compared to the control, and 31.35% and 27.71% when comparing the concentrations of 3% and 1%, respectively. Akhtar & Mahmood (1996) found satisfactory results when they used the commercial extract Nimin[®] at a concentration of 1 g L^{-1} to control *M. incognita* in pepper plants. Adegbite & Adesiyan (2005) also showed the efficiency of neem extract at a concentration of 0.25 g mL^{-1} , *in in vitro* tests, *in the* management of *M. incognita*. The satisfactory results found for Nematex applied at a concentration of 10% may have allowed it to remain in contact for longer and thus predisposed it to possibly act as a resistance inducer in a systemic way, through its application to the aerial part. Plant resistance against phytopathogens is induced when microbial components or products, using organic or inorganic compounds in contact with the plant (BONALDO et al., 2007), induce the plant to activate its structural and biochemical defense mechanisms in response to the presence of a pathogen (PASCHOLATI & TOFFANO, 2007).

The low stem development found in some plants exposes them to vulnerability to attack by pathogenic microorganisms and pests, compromising their performance. This presupposes that the extracts did not influence the increase in stem diameter, due to the non-release of the active ingredient, which may not have occurred due to the time the material was collected, the drying methods or the form of storage, since these factors can have a harmful influence on the preservation and effectiveness of the active ingredient (SCRAMIN et al., 1987).

Table 3. Stem diameter of pepper plants inoculated with *Meloidogyne Javanica,* treated with vegetable extracts in different concentrations and sources of application in greenhouse conditions, Bom Jesus-PI, 2012.

	Stem diameter								
	Forms of Application								
	Soil			Air Part			Soil + Air		
Extracts	1%	3%	10%	1%	3%	10%	1%	3%	10%
Witness	8.80 Aa	8.80 Aa	8.80 Aa	8.80 Aa	8.80 Aa	8.80 Ba	8.80 Aa	8.80 Aa	8.80 Aa
Rotenat	9.00 Aa	9.05 Aa	8.11 Aa	9.23 Aa	9.00 Aa	8.50 Ba	8.25 Ab	9.02 Ab	7.32 Ba
Pironat	8.15 Aa	8.36 Ab	8.40 Ab	8.29 Ab	8.79 Ab	8.22 Ba	8.95 Aa	8.35 Aa	8.51 Aa
Natunnem	8.51 Aa	8.86 Aa	8.53 Aa	8.50 Ab	7.91 Ab	5.23 Ca	8.40 Ab	8.32 Ab	6.93 Ba
Nematex	7.82 Aa	8.57 Aa	7.89 Aa	8.66 Aa	8.42 Aa	11.06 Ab	8.40 Aa	8.11 Aa	8.96 Aa

| Pyroligneous Acid | 8.05 Aa | 9.01 Aa | 8.36 Aa | 9.47 Ab | 8.12 Aa | 8.88 Bb | 8.44 Aa | 8.89 Aa | 9.19 Aa |

C. V. = 8.36

Averages followed by the same capital letter in the column and small letter in the row do not differ by the Scott-Knott test at 1% probability

For the dry phytomass of the aerial part (Table 4), there was no effect of the extracts depending on the different concentrations, or the results were lower when compared to the control, which suggests that the active ingredient of the extracts was not released, affecting the parasitism of the pepper plants and consequently compromising their development.

Table 4 - Aerial part dry phytomass of pepper plants inoculated with *Meloidogyne javanica*, treated with plant extracts in different concentrations under greenhouse conditions, Bom Jesus-PI, 2012.

Extracts	Aerial Part Dry Phytomass (g)		
	Concentrations		
	1%	3%	10%
Witness	23.43 Aa	23.43 Aa	23.43 Aa
Rotenat	25.41 Aa	23.05 Aa	21.41 Aa
Pironat	22.79 Aa	25.92 Aa	24.91 Aa
Natunnem	25.76 Ac	17.94 Bb	10.55 Ba
Nematex	24.79 Aa	23.85 Aa	26.05 Aa
Pyroligneous Acid	27.41 Aa	25.26 Aa	23.26 Aa

C. V. = 19.50

Averages followed by the same capital letter in the column and small letter in the row do not differ by the Scott-Knott test at 1% probability.

For the dry phytomass of the aerial part (Table 5), Pironat was the only extract that differed in terms of the increase in the aerial part of the treated plants, although the method of application did not influence the activity of the extracts. The other extracts did not differ significantly from the control for this parameter. The good results observed in the plants treated with Pironat can be attributed to its constituents, such as phenolic components, acids, alcohols, water and acetic acid which, when applied to the soil, improve its physical, chemical and biological properties, favoring the absorption of nutrients by the plants (ZANETTI, 2004).

Table 5 - Dry phytomass of the aerial part of pepper plants inoculated with *Meloidogyne javanica*, treated with plant extracts in different forms of application under greenhouse conditions, Bom Jesus-PI, 2012.

	Aerial Part Dry Phytomass (g)		
Forms of Application Extracts	Soil	Air Part	Soil + Air Part
Witness	23.43 Aa	23.43 Aa	23.43 Ba
Rotenat	22.83 Aa	26.00 Ab	21.04 Ba
Pironat	23.62 Aa	24.31 Aa	25.69 Aa
Natunnm	21.66 Ab	17.61 Ba	14.97 Ca
Nematex	24.25 Aa	26.21 Aa	24.22 Ba
Pyroligneous Acid	23.44 Aa	25.58 Aa	26.90 Aa

C. V. = 19.50

Averages followed by the same capital letter in the column and small letter in the row do not differ by the Scott-Knott test at 1% probability.

A well-developed root system is extremely important for the development of the plant, since diseases that compromise roots are among the main causes of reductions in the productivity of economically important crops (SAMI, 2005).

For the triple interaction of the root dry phytomass variable (Table 6), only the extracts Rotenat, Natunnem and Nematex, when applied to the soil, Pyroligneous Acid and Natunnem when applied in combination had a significant effect, so that the extracts interfered with parasitism, inhibiting its development and consequently enabling the development of the roots. The extracts applied only to the aerial part did not influence the increase in the plants' root system.

The products Natuneem® and Nematex® contain cold-extracted virgin neem oil, which is effective in managing phytopathogenic nematodes (NATURAL RURAL, 2010). Significant results with products of this nature have been gaining notoriety around the world. Biopirol® pyroligneous extract in nematode management, causing a reduction in the formation of galls on tomato roots was also found by Corbani (2008).

Research has shown satisfactory results when pyroligneous acid is applied for various purposes, such as pest and disease control (MIYASAKA et al., 2001), as an organic fertilizer (TSUZUKI et al., 2000), for medicinal use (LOO et al., 2008) and for human and animal nutrition (WANG et al., 2010). Ichikawa & Ota (1998), studying rice in the seedling stage, found that the application of pyroligneous extract to the soil caused greater development of the aerial part and root of the seedlings, and consequently improved development after transplanting.

According to Tsuzuki (1989), pyroligneous extract induces the development and elongation of new roots and the effect of the product is probably hormonal in nature. Root elongation is a response to microorganism attack, in which the plant stimulates the emission of healthy roots to contain the advance of the disease in the root system, favoring development and production (AGUIAR, 2012).

The promising results of these extracts demonstrate their efficiency in the management of both pests and diseases, which has already been proven in various studies as another simple and economically viable alternative for practical use in the control of phytopathogens (SOUZA et al., 2002; MORAIS, 2004, BASTOS & ALBUQUERQUE, 2004).

Commercial extracts are also used to manage agricultural pests (BONALDO et al. 2006). When evaluating the effect of natural products and thiamethoxam on thrips control in grapevines and their effects on natural enemies, Nali et al. (2004) found that Nim at 0.5% and Rotenat at 0.5% controlled 57.6% and 46.4% of thrips. Brito et al. (2006), evaluating the toxicity of Natuneem on *Tetranychus urticae* Koch and predatory mites of the phytoseiidae family, found that its toxic effect on eggs and their respective adults was higher than its toxicity to predatory mites.

Table 6. Root dry phytomass in pepper plants inoculated *with Meloidogyne Javanica,* treated with vegetable extracts at different concentrations and application sources in greenhouse conditions, Bom Jesus- PI, 2012.

Extracts	Root dry phytomass								
	Concentrations								
	Soil			Air Part			Soil + Air		
	1%	3%	10%	1%	3%	10%	1%	3%	10%
Witness	4.86 Ba	4.86 Aa	4.86 Aa	4.86 Aa	4.86 Aa	4.86 Aa	4.86 Ba	4.86 Aa	4.86 Aa
Rotenat	7.11 Aa	5.48 Aa	5.14 Aa	7.45 Aa	5.70 Aa	4.99 Aa	3.86 Ba	5.57 Aa	6.21 Aa
Pironat	4.03 Ba	5.67 Aa	5.00 Aa	4.97 Aa	5.71 Aa	4.48 Aa	4.13 Ba	5.89 Aa	5.23 Aa
Natunnem	6.00 Aa	4.72 Aa	4.56 Aa	5.64 Ab	3.12 Aa	2.00 Ba	6.89 Ab	5.03 Ab	1.01 Ba
Nematex	5.56 Aa	4.60 Aa	4.96 Aa	4.89 Aa	5.39 Aa	5.34 Aa	3.75 Ba	3.86 Aa	4.91 Aa
Pyroligneous Acid	3.55 Ba	6.88 Ab	4.59 Aa	6.17 Ab	3.86 Aa	6.69 Ab	5.69 Aa	5.92 Aa	7.21 Aa

C. V = 30.52

In Table 7, for the root volume parameter, the different concentrations of the extracts did not contribute to an increase in root volume, with some of the extracts showing lower results when compared to the control. The effect of the extracts may not have been demonstrated due to some environmental changes in the soil, such as porosity, organic matter content and temperature, all of which have been pointed out for decades (JENKINS & TAYLOR, 1967).

Table 7. Root volume of pepper plants inoculated with *Meloidogyne javanica*, treated with plant extracts in different concentrations under greenhouse conditions, Bom Jesus-PI, 2012.

| Extracts | Root volume (ml) | | |
| | Concentrations | | |
	1%	3%	10%
Witness	70.00 Aa	70.00 Aa	70.00 Aa

Rotenat	64.33 Aa	57.83 Aa	58.66 Aa
Pironat	44.58 Ba	69.41 Ab	59.33 Ab
Natunnem	68.75 Ac	49.83 Ab	33.25 Ba
Nematex	60.41 Aa	54.66 Aa	60.25 Aa
Pyroligneous Acid	55.91 Ba	63.00 Aa	62.50 Aa

C. V = 33.17

Averages followed by the same capital letter in the column and small letter in the row do not differ by the Scott-Knott test at 1% probability.

The reproductive characteristics were influenced by the factors studied, as well as in the triple interaction (Table 8), which demonstrates the interdependence of the factors, which better explains the results obtained.

Table 8. Summary of the analysis of variance (F values) of the factorial model for the sources of variation plant extracts (EXT), concentrations (C) and forms of application (FA) for nematological evaluations number of galls (NG), number of egg masses (NMO), *Meloidogyne javanica* (MJ), in the pepper crop, UFPI-CPCE- 2012.

Sources of variation	GL	NG	NMO	MJ [1]
Extracts	5	58,34**	46,12**	94,93**
Concentration	2	0,61 ns	0,91 ns	19,14**
Form of application	2	9,20**	5,72**	2,53 ns
Ext X C	10	5,97**	1,64 ns	11,69**
Ext X FA	10	0,90 ns	1,64 ns	7,81**
FA X C	4	2,88*	4,64**	3,11**
Interaction (Ext x C x FA)	20	5,71 **	3,58**	8,83**
Waste	159			
C.V (%)		21.18	19.75	21.54

* Significant at 5% probability; ** Significant at 1% probability; ns Not significant. C. V - coefficient of variation. [1] Original values transformed into square root of x + 0.5

For the number of galls variable (Table 9), the three-way interaction showed that there was no difference in relation to the concentrations, however, the ways in which the extracts were applied showed satisfactory results, differing statistically from the control, demonstrating a high potential for reducing the number of galls, with the pironat extract, which did not differ from the control. In the literature, researchers have also had success when applying extracts to control nematodes, both in the soil and on the aerial part (GARDIANO, 2008; MATEUS, 2012). When evaluating commercial extracts in the management of *M. javanica* in tomato via foliar spraying, Melo et al. (2012) obtained good results in reducing the number of galls and egg mass when using Neemseto and pyroligneous acid. Gardiano et al. (2009) found nematicidal activity in botanical extracts from different species when applied to tomato plants.

Neves et al. (2009) also obtained efficient results when evaluating the effect of the plant-based product Champon® on *M. javanica* in tomato plants, which reduced the number of galls and egg mass by 52.9% and 80.6%, respectively. Sobrinho (2008), when evaluating neem-based products in the control of some diseases of *Heliconia* spp. observed a 74.23% reduction in the population of the nematode *Helicotylenchus erythrinae* when he incorporated Neemtorta® into the soil.

For the egg mass variable in Table 10, although all the extracts differed from the control, the best results were obtained when Rotenat at 10% was applied to the soil and aerial part, Natunnem at 3% and 10% to the aerial part and combined, respectively, and pyroligneous acid at 3% when applied via foliar spraying. These results may have occurred in response to the attack of these pathogens by the plant, in which the plant activates some defense mechanism, induced by the extracts applied, since the active ingredients present in the extracts can act as elicitors and induce resistance in the host plants (VIDHYASKARAN, 1992).

The good results found in the neem-based product are due to the presence of the active ingredient azadirachtin, already observed by several researchers with promising results (NEVES et al., 2003, OLIVEIRA, 2005; JAVED, 2008). Azevedo et al. (2007), testing the efficiency of natural products such as Pironat® , Rotenat® , Natuneem® and other extracts in controlling *Callosobruchus maculatus* in stored cowpeas (*Vigna unguiculata* (L.)), found that these products caused mortality of 79%, 34% and 84% respectively.

It is important to note that the nematicidal or nematostatic effect of plant extracts sprayed on the aerial part of plants depends on the assimilation of bioactive compounds by the leaves, the translocation and release of these compounds by root exudates, and can also activate the induction of resistance in plants (CHITWOOD, 2002).

Table 9. Number of *Meloidogyne javanica* galls on pepper plants treated with vegetable extracts in different concentrations and forms of application under greenhouse conditions, Bom Jesus-PI, 2012.

	Number of Galls								
	Concentrations								
	Soil			Air Part			Soil + Air		
Extracts	1%	ooz 3 /0	10%	1%	ooz 3 ZO	10%	1%	ooz 3 ZO	10%
Witness	246.75 Ca	246.75 Ba	246.75 Ca	246.75 Ba	246.75 Da	246.75 Da	246.75 Ba	246.75 Ba	246.75 Ca
Rotenat	199.25 Cb	188.25 Ab	69.50 Aa	199.75 Ab	82.50 Aa	76.50 Aa	172.75 Aa	157.50 Aa	168.75 Ba
Pironat	85.00 Aa	162.50 Ab	198.25 Bb	101.25 Aa	178.50 Cb	87.50 Aa	154.50 Aa	132.50 Aa	179,00 Ba
Natunnem	147.25 Bb	163.50 Ab	105.75 Aa	89.00 Aa	81.00 Aa	171.25 Cb	160,00 Ab	141.75 Ab	87.50 Aa
Nematex	141.25 Ba	135.25 Aa	161.25 Ba	96.75 Aa	139.75 Ba	138.00 Ba	123.50 Aa	182.00 Aa	150,00 Ba
Acid Pyroligneous	162.25 Ba	191.25 Aa	129.50 Aa	203.25 Ba	178.50 Ca	179,00 Ca	165.50 Aa	166.75 Aa	199.25 Ba

C. V. = 21.18

Averages followed by the same capital letter in the column and small letter in the row do not differ by the Scott-Knott test at 1% probability.

Table 10. Number of egg masses of *Meloidogyne Javanica* in pepper plants treated with plant extracts in different concentrations and sources of application in greenhouse conditions, Bom Jesus-PI, 2012.

	Number of egg masses								
	Concentrations								
Extracts	Soil			Air Part			Soil + Air		
	1%	3%	10%	1%	3%	10%	1%	3%	10%

Treatment									
Witness	242.25 Ca	242.25 Ba	242.25 Ba	242.25 Ca	242.25 Da	242.25 Da	242.25 Ba	242.25 Ba	242.25 Ca
Rotenat	179.00 Bb	183.75 Ab	97.00 Ab	164.75 Aa	92.25 Aa	88.00 Ab	153.00 Aa	165.25 Aa	170.00 Bb
Pironat	163.25 Ba	145.00 Aa	155.25 Aa	132.00 Aa	169.25 Cb	133.25 Ba	136.25 Aa	182.25 Ab	192.50 Bb
Natunnem	171.50 Ba	178.25 Aa	143.00 Aa	134.25 Ab	102.25 Ab	187.00 Cb	152.25 Aa	194.25 Aa	97.25 Aa
Nematex	181.50 Ba	155.25 Aa	139.75 Aa	128.50 Aa	149.50 Ba	136.00 Ba	144.25 Aa	147.00 Aa	160.75 Ba
Acid Pyroligneous	113.50 Aa	184.50 Aa	135.25 Aa	191.25 Bb	102.00 Aa	180.75 Cb	166,00 Ab	169.50 Aa	184.25 Ba

C. V. = 19.75

Averages followed by the same capital letter in the column and small letter in the row do not differ by the Scott-Knott test at 1% probability.

For the variable number of *M. javanica* in the soil (Table 11), all the extracts were efficient, differing from the control, regardless of the form of concentration. These results demonstrate the effect of secondary metabolites in plants, with antimicrobial effects, produced in response to various weather conditions, which are capable of reducing the activity of pathogens in plant tissues (SCHWAN- ESTRADA et al., 2000).

Cuadra et al. (2000) found satisfactory results when using pyroligneous acid on juveniles *of M. incognita*, concluding that pyroligneous acid had nematicidal action. Gonçalves (2001) also showed the action of pyroligneous acid in controlling *M. incognita*, when used in a 1:50 dilution, reducing the nematode population when compared to the control.

The positive effects of neem extracts have been demonstrated in numerous studies. Javed et al. (2008), evaluating the effect of applying an aqueous neem extract, observed reductions in the number of galls and *M. javanica* eggs at doses of 1.5% and 3.0%. Almeida et al. (2012), when evaluating different ways of preparing plant extracts on *M. javanica* in tomato plants, observed a 92.1% reduction in the number of galls compared to the control.

The significant effect of the interaction demonstrated the activity of the compounds, as well as the activation of plant protection mechanisms.

Table 11. Number of *AQMeloidogyne Javanica* in the soil, in pepper plants treated with vegetable extracts, in different concentrations and sources of application in greenhouse conditions, Bom Jesus-PI, 2012.

Extracts	Soil			Air Part			Soil + Air		
	1%	3%	10%	1%	3%	10%	1%	3%	10%
Witness	415.00 Ba	415.00 Ba	415.00 Ba	415.00 Ba	415.00 Ba	415.00 Ba	415.00 Ba	415.00 Ba	415.00 Ba
Rotenat	107.50 Aa	167.50 Aa	100.00 Aa	107.50 Aa	95.00 Aa	90.00 Aa	167.50 Aa	120.00 Aa	187.50 Aa
Pironat	145.00 Aa	150.00 Aa	125.00 Aa	217.25 Aa	90.00 Aa	135.00 Aa	162.50 Aa	192.50 Aa	30.00 Aa
Natunnem	60.00 Aa	137.50 Aa	90.00 Aa	142.50 Aa	292.50 Ba	62.50 Aa	110.00 Aa	155.00 Aa	35.00 Aa
Nematex	45.00 Aa	175.50 Aa	80.00 Aa	40.00 Aa	62.50 Aa	42.50 Aa	80.00 Aa	47.50 Aa	60.00 Aa
Acid Pyroligneous	125.00 Aa	40.00 Aa	65.00 Aa	45.00 Aa	42.50 Aa	85.00 Aa	30.00 Aa	87.50 Aa	47.50 Aa

C. V. = 21.54

Means followed by the same capital letter in the column and small letter in the row do not differ by the Scott-Knott test at 1% probability

3.4 CONCLUSIONS

The extracts with commercial formulations were effective in managing phytonematoids.

All the extracts were effective in reducing the number of galls and juveniles in the soil, regardless of concentration.

The extracts Pironat, Natunnem and Nematex at 1% and Rotenat and Natunnem at 10% showed the best results in reducing the number of galls.

Rotenat at 10% applied to the soil and aerial part, Natunnem at 3% and 10% applied to the aerial part and combined and pyroligneous acid at 3% when applied via foliar spraying showed the best results in reducing egg masses.

3.5 BIBLIOGRAPHICAL REFERENCES

ADEGBITE, A.A.; ADESIYAN, S.O. Root extracts of plants to control root-knot nematode on edible soybean. **World Journal of Agricultural Sciences**, n. 1, v. 1, p. 18-21, 2005.

AGUIAR, C. **Botànica para Ciências Agràrias e do Ambiente**: Morfologia e funçao. Volume 1, Polytechnic Institute of Bragança, 2012, 123p.

AKHTAR, M.; MAHMOOD, I. Effect of a plant based product 'Nimin' and some plant oils on nematodes. **Nematologia Mediterranea**, v. 24, p. 3-5, 1996.

ALMEIDA[F.A.] ; PETTER, F.A.; SIQUEIRA, V.C.; ALCÂNTARA NETO; F;

ALVES, A.U.;LEITE,M.L.T. Modes of preparation of plant extracts on *Meloidogyne javanica* in tomato. **Nematropica**, n.1, v. 42, p. 9-15. 2012. AZEVEDO, F.R.; GUIMARÂES, J.A.; BRAGA SOBRINHO, R.; LIMA, M.A.A. Eficiência de produtos naturais para o controle de *bemisia tabaci* biótipo b (hemiptera: aleyrodidae) em meloeiro. **Arq. Instituto Biològico**, Sao Paulo, v.72, n.1, p.73-79, jan./mar., 2005.

AZEVEDO, F.R.; LEITÂO, A.C.L.; LIMA, M.A.A.; GUIMARÂES, J.A. Efficiency of natural products in the control of *Callosobruchus maculatus* (Fab) in cowpea (*Vigna unguiculata* (L.) Walp. **Revista Ciência Agronômica**, n.2, v.38, p.182-187, 2007.

BASTOS, C.N.; ALBUQUERQUE, P.S.B. Efeito de óleo de *Piper aduncum* no controle em pós colheita de *Colletotrichum musae* em banana. **Fitopatologia Brasileira**. n. 5, v. 29, p.555-557, 2004.

BONALDO, S.M.; SCHWAN-ESTRADA, K.R.F.; STANGARLIN, J.R.; CRUZ, M.E.S; FIORO-TUTIDA, A.C.G. Contribution to the study of antifungal and phytoalexin eliciting activities in sorghum and soybean by *eucalyptus* (*Eucalyptus citriodora*). **Summa Phytopathologica**, Botucatu, n. 3, v.35, p.383-387, 2007.

BRITO H. M.; GONDIM JUNIOR, M.G.C.; OLIVEIRA, J.V. CÂMARA, C.A.G.

Toxicity of Natuneem on *Tetranychus urticae* Koch (acari: Tetranychidae) and predatory mites of the phytoseiidae family. **Ciência agrotecnica**, Lavras, n. 4, v. 30,

p. 685-691, 2006.

CAMPOS, V.P. Diseases caused by nematodes in tomatoes. In: ZAMBOLIM, L., VALE, F.X.R.; COSTA, H. Contrôle de Doenças de Plantas - **Hortaliças**. UFV, Viçosa, p. 801-841. 2000.

CHITWOOD, D.J. Phytochemical based strategies for nematode control. **Annual Review of Phytopathology,** v. 40, p. 221-249, 2002.

COIMBRA J.L.; SOARES, A.C.F.; GARRIDO, M.S.; SOUSA, C.S.; RIBEIRO, F.L.B. Toxicity of plant extracts *to Scutellonema bradys*. **Pesquisa agropecuâria Brasileira**, v.41, p.1209-1211, 2006.

COOLEN, W.A.; D'HERDE, C.J. A method for the quantitative extraction of nematodes from plant tissue. Ghent, Belgian: **State of Nematology and Entomology Research Station**, 77 p, 1972.

CORBANI, R.Z. **Study of BIOPIROL® Pyroligneous Extract in the management of nematodes in sugar cane, olericolas and citrus, in different environments**. 55f. Thesis (Doctorate in Agronomy). Faculty of Agricultural and Veterinary Sciences. Paulista State University. Jaboticabal, 2008.

COSTA, M.J.N.; CAMPOS, V.P; OLIVEIRA, D.F.; BONETI, J.I.S.; FERRAZ, S. Modification of the Hussey and Barker method for extraction of *Meloidogyne exigua* eggs from coffee trees. **Fitopatologia Brasileira,** v. 6, p. 553, 1991.

CUADRA, R.; CRUZ, X.; PERERA, E.; MARTIN, E.; DIAZ, A. Algunos compuestos naturales com efecto nematicida.**Revista de Protección Vegetal**, La Habana, v.24, n.15, p.31-37, 2000.

DOIHARA, I.P.; **Effect of applying pyroligneous extract, neem oil (*Azadirachta indica*) and acibenzolar-s-methyl on nematode-host plant interaction**. RECIFE May 2005.

BRAZILIAN AGRICULTURAL RESEARCH company - EMBRAPA.

Gall-forming nematodes. Pelotas-RS. 2006.

ESECHIE, H. A. Assessment of pyroligneous liquid as a potential organic fertilizer. In: INTERNATIONAL CONFERENCE ON ECOLOGICAL AGRICULTURE: TOWARDS SUSTAINABLE DEVELOPMENT, Muscat. **Proceedings** Chandigarh: **Centre for Research in Rural and Industrial Development**, v.1, p.591-595, 1998.

FRANZENER, G.; MARTINES-FANZENER, A.S.; STANGARLIN, J.R.; FURLAMETTO, C. SCHWAN-ESTRADA, K.R.F. Protection of tomato plants against *Meloidogyne incognita* by the aqueous extract of *Taguetes patula*. **Nematologia Brasileira**. n.1, v.31, p. 27-36, 2007.

GALLO, D.; NAKANO, O.; SILVEIRA NETO, S.; CARVALHO, R.P.L.; BAPTISTA, G.C.; BERTI FILHO, E.; PARRA, R.P.; ZUCCHI, R.A.; ALVES, S.B.; VENDRAMIN, J. D.; MARCHINI, L.C.; LOPES, J.R.C.; OMOTO, C. **Entomologia Agricola.** Piracicaba: FEALQ, 2002. 920p.

GARDIANO, C.G. **The nematicidal activity of aqueous extracts and vegetable tinctures on *Meloidogyne javanica.*** 92f; Dissertation (Master's Degree) Federal University of Viçosa, 2006.

GARDIANO, C.G.; FERRAZ, S.; LOPES, E.A.; FERREIRA, P.A.; CARVALHO, S. L.; FREITAS, L.G. Evaluation of Aqueous Extracts of Plant Species, Applied via Foliar Spraying, on *Meloidogyne javanica. **Summa Phytopathologica***, Botucatu, n. 4, v. 34, p. 376-377, 2008.

GARDIANO, C.G., S. FERRAZ, E.A. LOPES, P.A. FERREIRA, D. X. AMORA, L.G. Evaluation of aqueous extracts of various plant species, applied to the soil, on *Meloidogyne javanica* (Treub, 1885) Chitwood, 1949. **Semina: Agricultural Sciences** v. 30, p. 551-556, 2009.

GONÇALVES, R.R. **The use of *pyroligneous* acid to control *Meloidogyne incognita* in tomato plants (*Lycopersicon esculentum*).** 2001. 24f Monograph - School of Agriculture and Sciences, Machado, 2001.

ICHIKAWA, T.; OTA, Y. Effect of pyroligneous acid on the growth of rice seedlings. **Japanese Journal of Crop Science,** Miyazaki, v.51, p.14-17, 1998.

JENKINS, W.R.; TAYLOR, D.P. Chemical control In. Plant nematology, Ed. Peter **Gray Reinhold publishing corporation**, New York, 222-223p, 1967.

JAVED, N.; GOWEN, S.R.; INAM-UL-HAQ, M.; ABDULLAH, K.; SHAHINA, F. effect of neem (Azadirachta indica) and persistent formulations against root-knot nematodes, Meloidogyne javanica and their storage life. **Crop Protection**, v. 26, p. 911-916, 2008.

KUROZAWA, C.; PAVAN, M.A.; KRAUSE-SAKATE, R. Diseases of solanaceous plants: eggplant, jiló, peppers and chili peppers. In: KIMATI, H.; AMORIM, L.; REZENDE, J.A.M.; FILHO BERGAMIN, A.; CAMARGO, L.E.A. **Manual de Fitopatologia: doenças das plantas cultivadas.** v.2. Sao Paulo: Editora Agronômica Ceres Ltda. 2005. p.589-596.

LOO A.Y. **Isolation and characterization of antioxidant compounds from pyroligneous acid of Rhizophora apiculata**. 2008. 239f. Thesis (Doctorate). Universiti Sains Malaysia. Penang, 2008.

LOPES, E. A.; FERRAZ, S.; FREITAS, L. G.; FERREIRA, P. A. Control of *Meloidogyne javanica* with different amounts of neem cake (Azadirachta indica). Revista Tròpica - **Ciências Agràrias e Biológicas**, n.1, v. 2, p. 17, 2008.

MIYASAKA, S.; OHKAWARA, T.; KUNIO, N. **Charcoal derivatives, pyroligneous extract and fine charcoal in natural agriculture:** handout. Miyazaki: APAN, 2001.

MOHAN, K. SUBHASHINI, S. Effect of organic amendments on parasitic nematodes of Okra (Abelmoschus esculentus L.). **Journal of Theoretical and Experimental Biology**, n.1, v.7, p. 53-55, 2010.

CONAB. Irrigation in pepper growing. 1st edition Brasilia, **Ministério da Agricultura Pecuària e Abastecimento**, 2012. (Circular tècnica 101).

MARTINEZ, M.M. Action of neem on nematodes. In: **Neem - *Azadirachta indica*, nature, multiplc uses and production**. IAPAR Instituto Agronòmico do Paranâ, Londrina-PR. p. 65-68, 2002.

MATEUS, M.A.F. Aqueous extracts of medicinal plants in the control of gall nematodes. 2012. 59f. Dissertation (Master's Degree) Midwestern State University.

MORANDI FILHO, W.J.; BOTTON, M.; GRÜTZMACHER, A.D.; GIOLO, F.P. MANZONI, C.G. Action of natural products on the survival of *Argyrotaenia sphaleropa* (Meyrick) (Lepidoptera: Tortricidae) and selectivity of insecticides used in organic grapevine production on *Trichogramma pretiosum* Riley (Hymenoptera: Trichogrammatidae). **Ciência Rural,** n. 4, v.36, p. 1072-1078, Santa Maria, July-August, 2006.

MOTOYAMA, M.M. SCHWAN-ESTRADA, K.R.F.; STANGARLIN, J.R.; FIORI-TUTIDA, A.C.G. SCAPIM, C.A. Induction of phytoalexins in soybeans and sorghum and fungitoxic effect of citrus extracts on Colletotrichum lagenarium and Fusarium semitectum. **Acta Sciencia Agronomica**, Maringà, v. 25, p. 491-496, 2003.

NALI, L.R.; BARBOSA, F.R.; CARVALHO, C.A.L.; SANTOS, J.B. C. Efficiency of natural insecticides and thiamethoxam in the control of thrips in grapevines and selectivity for natural enemies. **Pesticidas: Ecotoxicologica. e Meio Ambiente**, Curitiba, v. 14, p. 103-108, jan./dez.2004.

NATURAL RURAL. 2010. Available at: <http:// www.naturalrural.com.br>. Accessed on: 06/Feb./2013.

NEVES, B.P.; OLIVEIRA I. P.; NOGUEIRA, J.C.M. Cultivo e Utilização do Nim Indiano. **Empresa Brasileira de Pesquisa Agropecuâria**, Goiânia (GO), 12 p. (Technical circular, 62). 2003.

NEVES, E.J.M.; REISSMANN, C.B.; DEDECEK, R.A.; CARPANEZZI, A.A. Nutritional characterization of neem in plantations in Brazil. **Revista Brasileira de Engenharia Agricola e Ambiental,** n. 1, v.17, p.26-32, 2013.

NEVES, W.S.; FRREITAS, L.G.; COUTINHO, M.M.; DALLEMOLE-GIARETTA· R; FABRY, C.F.S; DHINGRA, O.D.; FERRAZ, S. Nematicidal action of extracts of garlic, mustard, chili pepper, mustard oil and two products based on capsinoids and allyl isothiocyanate on juveniles *of Meloidogyne javanica*, (treub) Chitwood, 1949, in a greenhouse. **Summa phytopathologica**, n. 4, v.35, Botucatu Oct/Dec. 2009.

OLIVEIRA, F.S.; ROCHA, M.R.; REIS, A.J.S.; MACHADO, V.O.F.; SOARES, R.A.B. Efeito de produtos quimicos e naturais sobre a populaçao de nematóide Pratylenchus brachyurus na cultura da cana-de-açùcar. **Pesquisa Agropecuâria Tropical**, Goiânia, v. 35, p. 171-178, 2005.

OOSTENBRINK, M. Major characteristics of the relation between nematodes and plants. **Meded. Landbouw**, Wageningen, p. 66, 1966.

PASCHOLATI, S.F.; TOFFANO, L. Induction of resistance against phytopathogens in tree species. In: Rodrigues, F.A.; Romeiro, R.S. **Induçâo de resistência em plantas a patógenos**. Viçosa: Universidade Federal de Viçosa, v.1, cap.3, p.5966, 2007.

RODRIGUES, F.; DATNOFF, L.E; KORNDORFER, G.H.; SEEBOLD, K.W.; RUSH, M.C. Effect of silicon and host resistance on sheath blight development in rice. **Plant Disease**, n. 8, v.85, p.827-832, 2001.

SAMI, J. M. Ecology and management of root pathogens in tropical soils. Recife: UFRPE, University Press, 2005. 398 p.

SANTOS, P.V. Reaçao de acessos de pimenteiras (*Capsicum* spp.) a Meloidogyne incognita RAÇA 3. 2008, 96f. Dissertaçao (Master's Degree) Universidade Estadual de Santa Cruz, BA.

SCHMUTTERER, H. Properties and potential of natural pesticides from the neem tree, *Azadirachta indica*. **Revista Entomologia**, v. 35, p. 271-297, 1990.

SCRAMIN, S.; FERNANDES, L.M.S.; SILVA, H.P.; YAHN, C. Nematicidal activity of plant extracts on *Meloidogyne incognita*. **Fitopatologia Brasileira**, n.2, v.12, p.151, 1987.

SIVAKUMAR, M. GUNASEKARAN, K. Management of root-knot nematodes in tomato, chilli and brinjal by neem oil formulations. **Journal of Biopesticides**, n. 2, v.4, p.198-200, 2011.

SOBRINHO, C.C.M. **Phytosanitary diagnosis and evaluation of neem in the control of some pests of *Heliconia* spp. on the south coast of Bahia**. 96p; Dissertation (Master's Degree) Universidade Estadual de Santa Cruz, Ilhéus, BA, 2008.

SOUZA, M.A.A.; BORGES, R.S.O.S.; STARK, M L.M.; SOUZA, S.R. Effect of aqueous, methanolic and ethanolic extracts of medicinal plants on the germination of lettuce seeds and on the mycelial development of phytopathogenic fungi of agricultural interest. **Revista Universidade Rural** v. 22, p.181185, 2002.

TAYLOR, A.L.; SASSER, J.N. Biology, identification and control of root-knot nematodes (Meloidogyne species). Raleigh: **International Meloidogyne Project**, NCSU & USAID Coop. Publ., 1978. 111p.

TSUZUKI, E. Effect of Pyroligneous Acid and Mixture of Charcoal with Pyroligneous Acid on the Growth and Yield of Rice Plant . **Japan Journal Crop Science,** Miyazaki, v.58, p.592-597, 1989.

TSUZUKI, E.; MORIMITSU, T.; MATSUI, T.; Effects of chemical compounds in pyroligneous acid on root rice plant. **Japan Journal Crop Science**, Tokyo, n. 4, v. 66, p. 15-16, 2000.

WANG, Z.; LIN, W.; SONG, W.; YAO, J. Preliminary investigation on concentrating of acetol from wood vinegar. **Energy Conversion and Management**, Belton, n. 2, v. 51, p.346-349, 2010.

VIDHYASKARAN, P. Principles of Plant Pathology. New Delhi: **CBS Printer and Publishers.** 1992.

WILCKEN, S.R.S.; GARCIA, M.J.; SILVA, N. Resistance of American-type lettuce to *Meloidogyne. Incognita* Race 2. **Nematologia Brasileira,** Piracicaba, v. 29, n. 2 p. 267- 271. 2005.

ZANETTI, M. **Use of by-products from the manufacture of charcoal in the formation of Iimoeiro 'Cravo' rootstock in a protected environment**. 77f. Dissertation (Master's Degree in Plant Production) - Universidade Estadual Paulista, Faculdade de Ciências Agràrias e Veterinàrias, Jaboticabal, 2004.

Printed by Books on Demand GmbH, Norderstedt / Germany